50% OFF Online GED Prep Course!

Dear Customer,

We consider it an honor and a privilege that you chose our GED Study Guide. As a way of showing our appreciation and to help us better serve you, we have partnered with Mometrix Test Preparation to offer **50% off their online GED Prep Course.** Many GED courses are needlessly expensive and don't deliver enough value. With their course, you get access to the best GED prep material, and you only pay half price.

Mometrix has structured their online course to perfectly complement your printed study guide. The GED Prep Course contains **in-depth lessons** that cover all the most important topics, **320+ video reviews** that explain difficult concepts, **over 1,250 practice questions** to ensure you feel prepared, and **430 digital flashcards**, so you can fit some studying in while you're on the go.

Online GED Prep Course

Topics Covered:

- Reasoning Through Language Arts
  - Reading for Meaning
  - Identifying and Creating Arguments
  - Agreement and Sentence Structure
- Mathematical Reasoning
  - Geometry
  - Statistics and Probability
  - Algebra
- Science
  - Designing and Interpreting Science Experiments
  - Using Numbers and Graphics in Science
  - Science Knowledge Overview
- Social Studies
  - Analyzing Historical Events and Arguments in Social Studies
  - Social Studies Knowledge Overview

Course Features:

- GED Study Guide
  - Get content that complements our best-selling study guide.
- 3 Full-Length Practice Tests
  - With over 1,250 practice questions, you can test yourself again and again.
- Mobile Friendly
  - If you need to study on-the-go, the course is easily accessible from your mobile device.
- GED Flashcards
  - The course includes a flashcard mode consisting of over 430 content cards to help you study.

To receive this discount, simply head to their website: mometrix.com/university/ged or simply scan this QR code with your smartphone. At the checkout page, enter the discount code: **TPBGED50**

If you have any questions or concerns, please don't hesitate to contact Mometrix at universityhelp@mometrix.com.

Sincerely,

in partnership with

FREE Test Taking Tips Video/DVD Offer

To better serve you, we created videos covering test taking tips that we want to give you for FREE. **These videos cover world-class tips that will help you succeed on your test.**

We just ask that you send us feedback about this product. Please let us know what you thought about it—whether good, bad, or indifferent.

To get your **FREE videos**, you can use the QR code below or email freevideos@studyguideteam.com with "Free Videos" in the subject line and the following information in the body of the email:

a. The title of your product

b. Your product rating on a scale of 1-5, with 5 being the highest

c. Your feedback about the product

If you have any questions or concerns, please don't hesitate to contact us at info@studyguideteam.com.

Thank you!

GED Math Study Guide 2023-2024

3 Practice Exams and GED Test Prep Book

[6th Edition]

Joshua Rueda

Written and edited by TPB Publishing.

ISBN 13: 9781637759899
ISBN 10: 1637759894

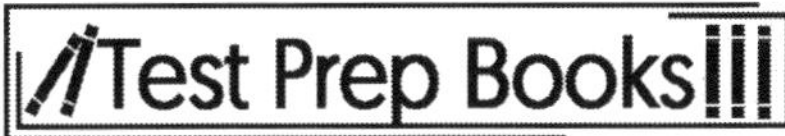

Table of Contents

Welcome

Dear Reader,

Welcome to your new Test Prep Books study guide! We are pleased that you chose us to help you prepare for your exam. There are many study options to choose from, and we appreciate you choosing us. Studying can be a daunting task, but we have designed a smart, effective study guide to help prepare you for what lies ahead.

Whether you're a parent helping your child learn and grow, a high school student working hard to get into your dream college, or a nursing student studying for a complex exam, we want to help give you the tools you need to succeed. We hope this study guide gives you the skills and the confidence to thrive, and we can't thank you enough for allowing us to be part of your journey.

In an effort to continue to improve our products, we welcome feedback from our customers. We look forward to hearing from you. Suggestions, success stories, and criticisms can all be communicated by emailing us at info@studyguideteam.com.

Sincerely,
Test Prep Books Team

FREE Videos/DVD OFFER

Doing well on your exam requires both knowing the test content and understanding how to use that knowledge to do well on the test. We offer completely FREE test taking tip videos. ***These videos cover world-class tips that you can use to succeed on your test.***

To get your ***FREE videos****, you can use the QR code below or email freevideos@studyguideteam.com with "Free Videos" in the subject line and the following information in the body of the email:*

a. The title of your product
b. Your product rating on a scale of 1-5, with 5 being the highest
c. Your feedback about the product

If you have any questions or concerns, please don't hesitate to contact us at info@studyguideteam.com.

Quick Overview

As you draw closer to taking your exam, effective preparation becomes more and more important. Thankfully, you have this study guide to help you get ready. Use this guide to help keep your studying on track and refer to it often.

This study guide contains several key sections that will help you be successful on your exam. The guide contains tips for what you should do the night before and the day of the test. Also included are test-taking tips. Knowing the right information is not always enough. Many well-prepared test takers struggle with exams. These tips will help equip you to accurately read, assess, and answer test questions.

A large part of the guide is devoted to showing you what content to expect on the exam and to helping you better understand that content. In this guide are practice test questions so that you can see how well you have grasped the content. Then, answer explanations are provided so that you can understand why you missed certain questions.

Don't try to cram the night before you take your exam. This is not a wise strategy for a few reasons. First, your retention of the information will be low. Your time would be better used by reviewing information you already know rather than trying to learn a lot of new information. Second, you will likely become stressed as you try to gain a large amount of knowledge in a short amount of time. Third, you will be depriving yourself of sleep. So be sure to go to bed at a reasonable time the night before. Being well-rested helps you focus and remain calm.

Be sure to eat a substantial breakfast the morning of the exam. If you are taking the exam in the afternoon, be sure to have a good lunch as well. Being hungry is distracting and can make it difficult to focus. You have hopefully spent lots of time preparing for the exam. Don't let an empty stomach get in the way of success!

When travelling to the testing center, leave earlier than needed. That way, you have a buffer in case you experience any delays. This will help you remain calm and will keep you from missing your appointment time at the testing center.

Be sure to pace yourself during the exam. Don't try to rush through the exam. There is no need to risk performing poorly on the exam just so you can leave the testing center early. Allow yourself to use all of the allotted time if needed.

Remain positive while taking the exam even if you feel like you are performing poorly. Thinking about the content you should have mastered will not help you perform better on the exam.

Once the exam is complete, take some time to relax. Even if you feel that you need to take the exam again, you will be well served by some down time before you begin studying again. It's often easier to convince yourself to study if you know that it will come with a reward!

Test-Taking Strategies

1. Predicting the Answer

When you feel confident in your preparation for a multiple-choice test, try predicting the answer before reading the answer choices. This is especially useful on questions that test objective factual knowledge. By predicting the answer before reading the available choices, you eliminate the possibility that you will be distracted or led astray by an incorrect answer choice. You will feel more confident in your selection if you read the question, predict the answer, and then find your prediction among the answer choices. After using this strategy, be sure to still read all of the answer choices carefully and completely. If you feel unprepared, you should not attempt to predict the answers. This would be a waste of time and an opportunity for your mind to wander in the wrong direction.

2. Reading the Whole Question

Too often, test takers scan a multiple-choice question, recognize a few familiar words, and immediately jump to the answer choices. Test authors are aware of this common impatience, and they will sometimes prey upon it. For instance, a test author might subtly turn the question into a negative, or he or she might redirect the focus of the question right at the end. The only way to avoid falling into these traps is to read the entirety of the question carefully before reading the answer choices.

3. Looking for Wrong Answers

Long and complicated multiple-choice questions can be intimidating. One way to simplify a difficult multiple-choice question is to eliminate all of the answer choices that are clearly wrong. In most sets of answers, there will be at least one selection that can be dismissed right away. If the test is administered on paper, the test taker could draw a line through it to indicate that it may be ignored; otherwise, the test taker will have to perform this operation mentally or on scratch paper. In either case, once the obviously incorrect answers have been eliminated, the remaining choices may be considered. Sometimes identifying the clearly wrong answers will give the test taker some information about the correct answer. For instance, if one of the remaining answer choices is a direct opposite of one of the eliminated answer choices, it may well be the correct answer. The opposite of obviously wrong is obviously right! Of course, this is not always the case. Some answers are obviously incorrect simply because they are irrelevant to the question being asked. Still, identifying and eliminating some incorrect answer choices is a good way to simplify a multiple-choice question.

4. Don't Overanalyze

Anxious test takers often overanalyze questions. When you are nervous, your brain will often run wild, causing you to make associations and discover clues that don't actually exist. If you feel that this may be a problem for you, do whatever you can to slow down during the test. Try taking a deep breath or counting to ten. As you read and consider the question, restrict yourself to the particular words used by the author. Avoid thought tangents about what the author *really* meant, or what he or she was *trying* to say. The only things that matter on a multiple-choice test are the words that are actually in the question. You must avoid reading too much into a multiple-choice question, or supposing that the writer meant something other than what he or she wrote.

5. No Need for Panic

It is wise to learn as many strategies as possible before taking a multiple-choice test, but it is likely that you will come across a few questions for which you simply don't know the answer. In this situation, avoid panicking. Because most multiple-choice tests include dozens of questions, the relative value of a single wrong answer is small. As much as possible, you should compartmentalize each question on a multiple-choice test. In other words, you should not allow your feelings about one question to affect your success on the others. When you find a question that you either don't understand or don't know how to answer, just take a deep breath and do your best. Read the entire question slowly and carefully. Try rephrasing the question a couple of different ways. Then, read all of the answer choices carefully. After eliminating obviously wrong answers, make a selection and move on to the next question.

6. Confusing Answer Choices

When working on a difficult multiple-choice question, there may be a tendency to focus on the answer choices that are the easiest to understand. Many people, whether consciously or not, gravitate to the answer choices that require the least concentration, knowledge, and memory. This is a mistake. When you come across an answer choice that is confusing, you should give it extra attention. A question might be confusing because you do not know the subject matter to which it refers. If this is the case, don't eliminate the answer before you have affirmatively settled on another. When you come across an answer choice of this type, set it aside as you look at the remaining choices. If you can confidently assert that one of the other choices is correct, you can leave the confusing answer aside. Otherwise, you will need to take a moment to try to better understand the confusing answer choice. Rephrasing is one way to tease out the sense of a confusing answer choice.

7. Your First Instinct

Many people struggle with multiple-choice tests because they overthink the questions. If you have studied sufficiently for the test, you should be prepared to trust your first instinct once you have carefully and completely read the question and all of the answer choices. There is a great deal of research suggesting that the mind can come to the correct conclusion very quickly once it has obtained all of the relevant information. At times, it may seem to you as if your intuition is working faster even than your reasoning mind. This may in fact be true. The knowledge you obtain while studying may be retrieved from your subconscious before you have a chance to work out the associations that support it. Verify your instinct by working out the reasons that it should be trusted.

8. Key Words

Many test takers struggle with multiple-choice questions because they have poor reading comprehension skills. Quickly reading and understanding a multiple-choice question requires a mixture of skill and experience. To help with this, try jotting down a few key words and phrases on a piece of scrap paper. Doing this concentrates the process of reading and forces the mind to weigh the relative importance of the question's parts. In selecting words and phrases to write down, the test taker thinks about the question more deeply and carefully. This is especially true for multiple-choice questions that are preceded by a long prompt.

9. Subtle Negatives

One of the oldest tricks in the multiple-choice test writer's book is to subtly reverse the meaning of a question with a word like *not* or *except*. If you are not paying attention to each word in the question, you can easily be led astray by this trick. For instance, a common question format is, "Which of the following is...?" Obviously, if the question instead is, "Which of the following is not...?," then the answer will be quite different. Even worse, the test makers are aware of the potential for this mistake and will include one answer choice that would be correct if the question were not negated or reversed. A test taker who misses the reversal will find what he or she believes to be a correct answer and will be so confident that he or she will fail to reread the question and discover the original error. The only way to avoid this is to practice a wide variety of multiple-choice questions and to pay close attention to each and every word.

10. Reading Every Answer Choice

It may seem obvious, but you should always read every one of the answer choices! Too many test takers fall into the habit of scanning the question and assuming that they understand the question because they recognize a few key words. From there, they pick the first answer choice that answers the question they believe they have read. Test takers who read all of the answer choices might discover that one of the latter answer choices is actually *more* correct. Moreover, reading all of the answer choices can remind you of facts related to the question that can help you arrive at the correct answer. Sometimes, a misstatement or incorrect detail in one of the latter answer choices will trigger your memory of the subject and will enable you to find the right answer. Failing to read all of the answer choices is like not reading all of the items on a restaurant menu: you might miss out on the perfect choice.

11. Spot the Hedges

One of the keys to success on multiple-choice tests is paying close attention to every word. This is never truer than with words like almost, most, some, and sometimes. These words are called "hedges" because they indicate that a statement is not totally true or not true in every place and time. An absolute statement will contain no hedges, but in many subjects, the answers are not always straightforward or absolute. There are always exceptions to the rules in these subjects. For this reason, you should favor those multiple-choice questions that contain hedging language. The presence of qualifying words indicates that the author is taking special care with their words, which is certainly important when composing the right answer. After all, there are many ways to be wrong, but there is only one way to be right! For this reason, it is wise to avoid answers that are absolute when taking a multiple-choice test. An absolute answer is one that says things are either all one way or all another. They often include words like *every*, *always*, *best*, and *never*. If you are taking a multiple-choice test in a subject that doesn't lend itself to absolute answers, be on your guard if you see any of these words.

12. Long Answers

In many subject areas, the answers are not simple. As already mentioned, the right answer often requires hedges. Another common feature of the answers to a complex or subjective question are qualifying clauses, which are groups of words that subtly modify the meaning of the sentence. If the question or answer choice describes a rule to which there are exceptions or the subject matter is complicated, ambiguous, or confusing, the correct answer will require many words in order to be expressed clearly and accurately. In essence, you should not be deterred by answer choices that seem excessively long. Oftentimes, the author of the text will not be able to write the correct answer without offering some qualifications and modifications. Your job is to read the answer choices thoroughly and completely and to select the one that most accurately and precisely answers the question.

13. Restating to Understand

Sometimes, a question on a multiple-choice test is difficult not because of what it asks but because of how it is written. If this is the case, restate the question or answer choice in different words. This process serves a couple of important purposes. First, it forces you to concentrate on the core of the question. In order to rephrase the question accurately, you have to understand it well. Rephrasing the question will concentrate your mind on the key words and ideas. Second, it will present the information to your mind in a fresh way. This process may trigger your memory and render some useful scrap of information picked up while studying.

14. True Statements

Sometimes an answer choice will be true in itself, but it does not answer the question. This is one of the main reasons why it is essential to read the question carefully and completely before proceeding to the answer choices. Too often, test takers skip ahead to the answer choices and look for true statements. Having found one of these, they are content to select it without reference to the question above. Obviously, this provides an easy way for test makers to play tricks. The savvy test taker will always read the entire question before turning to the answer choices. Then, having settled on a correct answer choice, he or she will refer to the original question and ensure that the selected answer is relevant. The mistake of choosing a correct-but-irrelevant answer choice is especially common on questions related to specific pieces of objective knowledge. A prepared test taker will have a wealth of factual knowledge at their disposal, and should not be careless in its application.

15. No Patterns

One of the more dangerous ideas that circulates about multiple-choice tests is that the correct answers tend to fall into patterns. These erroneous ideas range from a belief that B and C are the most common right answers, to the idea that an unprepared test-taker should answer "A-B-A-C-A-D-A-B-A." It cannot be emphasized enough that pattern-seeking of this type is exactly the WRONG way to approach a multiple-choice test. To begin with, it is highly unlikely that the test maker will plot the correct answers according to some predetermined pattern. The questions are scrambled and delivered in a random order. Furthermore, even if the test maker was following a pattern in the assignation of correct answers, there is no reason why the test taker would know which pattern he or she was using. Any attempt to discern a pattern in the answer choices is a waste of time and a distraction from the real work of taking the test. A test taker would be much better served by extra preparation before the test than by reliance on a pattern in the answers.

Introduction to the GED

Function of the Test

The General Education Development (GED) test is an exam developed and administered by the GED Testing Service, a joint venture of the American Council on Education and Pearson VUE. The GED offers those without a high school diploma the chance to earn a high school equivalency credential by evaluating their knowledge of core high school subjects.

GED test takers represent a wide age group with diverse goals. Generally, the GED is appropriate for people who did not graduate from high school but who wish to pursue advancement in their career and/or education. According to MyGED, approximately 98% of U.S. colleges and universities accept a GED as the equivalent of a high school diploma (other schools may require additional preparation courses in addition to a passing GED score in order to be considered for admission). Over 20 million adults have earned GED credentials, and the latest reported pass rates for the 2014 GED are around 60%.

Test Administration

GED tests are widely offered throughout the United States and Canada, although jurisdictions (state, province, etc.) may vary in terms of things like pricing, scheduling, and test rules. For international students and US military, international testing options are also available. Official GED Testing Centers are often operated by community colleges, adult education centers, and local school boards; GED Testing Service offers a comprehensive search of nearby test centers.

Keeping in mind that rules may vary between jurisdictions, all tests are administered in-person and taken on a computer. Tests are scheduled throughout the year; candidates should refer to their local testing centers for available test times. Candidates may take one of the four subject tests in an administration, or multiple (up to all four). Generally, test takers are able to take any test module three times without any restrictions on retesting. However, after three failed attempts, the candidate must wait a minimum of 60 days to retake the test. GED testing centers can also offer accommodations for students with disabilities, such as additional test time or Braille format tests. Test takers can request these accommodations when they register for an account on GED.com; approvals occur on an individual basis and typically take 30 days to receive.

Test Format

The GED consists of four sections, or modules: Mathematical Reasoning, Science, Social Studies, and Reasoning Through Language Arts. As mentioned, although the complete test is offered together, it is not necessary to take all four modules on one day. Note that for the Mathematical Reasoning section, a formula sheet will be provided. Test subjects vary in length. The following chart provides information about the sections:

Subject	Time	Topics
Mathematical Reasoning	115 minutes	Basic Math, Geometry, Basic Algebra, Graphs and Functions
Science	90 minutes	Reading for Meaning, Designing and Interpreting Science Experiments, Using Numbers and Graphs in Science
Social Studies	70 minutes	Reading for Meaning, Analyzing Historical Events and Arguments, Using Numbers and Graphs in Social Studies
Reasoning Through Language Arts	150 minutes	Reading for Meaning, Identifying and Creating Arguments, Grammar and Language

A ten-minute break is given between each module.

On the testing day, test takers are not permitted to eat, drink, smoke, or use their cell phones during the test. Test takers are permitted to bring a handheld calculator (TI-30XS Multiview Scientific Calculator) to the test; testing centers will not provide handheld calculators, although an on-screen calculator will be available on the computer. Students will also be provided with three erasable note boards to use during the test.

Scoring

Because the GED is now a computer-based test, scores will be available on MyGED within 24 hours of completing the test. The four modules of the GED are scored on a scale of 100–200. In order to earn high school equivalency, it is necessary to achieve a passing score on all of the four modules, and scores cannot be made up between modules—that is, a high score on one subject cannot be used to compensate for a low score on another subject. Scores are divided into four ranges:

1. A score lower than 145 points earns a score of "Not Passing." It is necessary to retake the test to earn high school equivalency.

2. A score at or higher than 145 points earns "GED Passing Score/High School Equivalency."

3. A score of 165-175 is deemed "GED College Ready." This designation advises colleges and universities that the test taker is ready to begin a degree program without further placement testing or preparation courses (policies vary among schools).

4. A score over 175 earns the test taker "GED College Ready + Credit." For some institutions, a score at this level allows the GED graduate to earn college credit for certain courses (policies vary among schools).

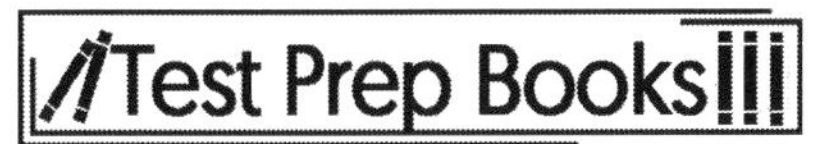

Test takers are encouraged to check what scores are required for admission by the colleges they intend to apply to, as some schools may seek scores that differ than the typical requisite 145. In 2017, the average scores of all test takers (including those who did not pass) were 154 for Science, 153 for Social Studies, 152 for Reasoning in Language Arts, and 150 for Math.

Study Prep Plan for the GED Math Exam

1 **Schedule -** Use one of our study schedules below or come up with one of your own.

2 **Relax -** Test anxiety can hurt even the best students. There are many ways to reduce stress. Find the one that works best for you.

3 **Execute -** Once you have a good plan in place, be sure to stick to it.

One Week Study Schedule

Day 1	Mathematical Reasoning
Day 2	Geometry
Day 3	Basic Algebra
Day 4	Graphs and Functions
Day 5	GED Mathematical Reasoning Practice Test #1
Day 6	GED Mathematical Reasoning Practice Test #3
Day 7	Take Your Exam!

Two Week Study Schedule

Day 1	Mathematical Reasoning	Day 8	Practice Test #1
Day 2	Whole Numbers, Fractions, and Decimal...	Day 9	Answer Explanations #1
Day 3	Geometry	Day 10	Practice Test #2
Day 4	Graphical Data Including Graphs, Tables, and...	Day 11	Answer Explanations #2
Day 5	Basic Algebra	Day 12	Practice Test #3
Day 6	Writing an Expression from a Written...	Day 13	Answer Explanations #3
Day 7	Graphs and Functions	Day 14	Take Your Exam!

One Month Study Schedule

Day 1	Mathematical Reasoning	Day 11	Mean, Median, Mode, and Range	Day 21	Features of Graphs and Tables for Linear and...
Day 2	Multiples and Factors	Day 12	Probability of an Event	Day 22	Equation of a Line from the Slope and a Point...
Day 3	Whole Numbers, Fractions, and Dec...	Day 13	Basic Algebra	Day 23	Functions in Tables and Graphs
Day 4	Properties of Operations	Day 14	Adding, Subtracting, Multiplying, Dividing...	Day 24	Practice Test #1
Day 5	Squares, Square Roots, Cubes, and Cube Roots	Day 15	Adding, Subtracting, Multiplying, Dividing...	Day 25	Answer Explanations for Practice Test #1
Day 6	Multiple-Step Problems that Use Ratios, Proportions...	Day 16	Solving a System of Two Linear Equations	Day 26	Practice Test #2
Day 7	Geometry	Day 17	Solving Inequalities and Graphing the Answer...	Day 27	Answer Explanations for Practice Test #2
Day 8	Area and Perimeter of Two-Dimensional Shapes	Day 18	Quadratic Equations with One Variable	Day 28	Practice Test #3
Day 9	Volume and Surface Area of Three...	Day 19	Graphs and Functions	Day 29	Answer Explanations for Practice Test #3
Day 10	Graphical Data Including Graphs...	Day 20	Proportional Relationships for...	Day 30	Take Your Exam!

Build your own prep plan by visiting:
testprepbooks.com/prep

Math Reference Sheet

Symbol	Phrase
+	added to, increased by, sum of, more than
-	decreased by, difference between, less than, take away
×	multiplied by, 3 (4, 5 . . .) times as large, product of
÷	divided by, quotient of, half (third, etc.) of
=	is, the same as, results in, as much as
x, *t*, *n*, etc.	a variable which is an unknown value or quantity
<	is under, is below, smaller than, beneath
>	is above, is over, bigger than, exceeds
≤	no more than, at most, maximum; less than or equal to
≥	no less than, at least, minimum; greater than or equal to
√	square root of, exponent divided by 2

Geometry	Description
$P = 2l + 2w$	for perimeter of a rectangle
$P = 4 \times s$	for perimeter of a square
$P = a + b + c$	for perimeter of a triangle
$A = \frac{1}{2} \times b \times h = \frac{bh}{2}$	for area of a triangle
$A = b \times h$	for area of a parallelogram
$A = \frac{1}{2} \times h(b_1 + b_2)$	for area of a trapezoid
$A = \frac{1}{2} \times a \times P$	for area of a regular polygon
$C = 2 \times \pi \times r$	for circumference (perimeter) of a circle
$A = \pi \times r^2$	for area of a circle
$c^2 = a^2 + b^2; c = \sqrt{a^2 + b^2}$	for finding the hypotenuse of a right triangle
$SA = 2xy + 2yz + 2xz$	for finding surface area
$V = \frac{1}{3}xyh$	for finding volume of a rectangular pyramid
$V = \frac{4}{3}\pi r^3; \frac{1}{3}\pi r^2 h; \pi r^2 h$	for volume of a sphere; a cone; and a cylinder

Radical Expressions	Description
$\sqrt[n]{a} = a^{\frac{1}{n}}, \sqrt[n]{a^m} = (\sqrt[n]{a})^m = a^{\frac{m}{n}}$	*a* is the radicand, *n* is the index, m is the exponent
$\sqrt{x^2} = (x^2)^{\frac{1}{2}} = x$	to convert square root to exponent
$a^m \times a^n = a^{m+n}$	multiplying radicands with exponents
$(a^m)^n = a^{m \times n}$	multiplying exponents
$(a \times b)^m = a^m \times b^m$	parentheses with exponents

Property	Addition	Multiplication
Commutative	$a + b = b + a$	$a \times b = b \times a$
Associative	$(a + b) + c = a + (b + c)$	$(a \times b) \times c = a \times (b \times c)$
Identity	$a + 0 = a;\ 0 + a = a$	$a \times 1 = a;\ 1 \times a = a$
Inverse	$a + (-a) = 0$	$a \times \frac{1}{a} = 1;\ a \neq 0$
Distributive	$a(b + c) = ab + ac$	

Data	Description
Mean	equal to the total of the values of a data set, divided by the number of elements in the data set
Median	middle value in an odd number of ordered values of a data set, or the mean of the two middle values in an even number of ordered values in a data set
Mode	the value that appears most often
Range	the difference between the highest and the lowest values in the set

Graphing	Description
(x, y)	ordered pair, plot points in a graph
$y = mx + b$	slope-intercept form; *m* represents the slope of the line and *b* represents the *y*-intercept
$f(x)$	read as *f* of *x*, which means it is a function of *x*
(x_2, y_2) and (x_2, y_2)	two ordered pairs used to determine the slope of a line
$m = \frac{y_2 - y_1}{x_2 - x_1}$	to find the slope of the line, *m*, for ordered pairs
$Ax + By = C$	standard form of an equation, also for solving a system of equations through the elimination method
$M = (\frac{x_1 + x_2}{2}, \frac{y_1 + y_2}{2})$	for finding the midpoint of an ordered pair
$y = ax^2 + bx + c$	quadratic function for a parabola
$y = a(x - h)^2 + k$	quadratic function for a parabola with vertex
$y = ab^x; y = a \times b^x$	function for exponential curve
$y = ax^2 + bx + c$	standard form of a quadratic function
$x = \frac{-b}{2a}$	for finding axis of symmetry in a parabola; given quadratic formula in standard form
$f = \sqrt{\frac{\Sigma(x - \bar{x})^2}{n - 1}}$	function for standard deviation of the sample; where $\bar{x}$ = sample mean and *n* = sample size

Proportions and Percentage	Description
$\frac{\text{gallons}}{\text{cost}} = \frac{\text{gallons}}{\text{cost}} : \frac{7 \text{ gallons}}{\$14.70} = \frac{x}{\$20}$	written as equal ratios with a variable representing the missing quantity
$\frac{y_1}{x_1} = \frac{y_2}{x_2}$	for direct proportions
$(y_1)(x_1) = (y_2)(x_2)$	for indirect proportions
$\frac{\text{change}}{\text{original value}} \times 100 = \text{percent change}$	for finding percentage change in value
$\frac{\text{new quantity} - \text{old quantity}}{\text{old quantity}} \times 100$	for calculating the increase or decrease in percentage

Mathematical Reasoning

Basic Math

Numbers usually serve as an adjective representing a quantity of objects. They function as placeholders for a value. Numbers can be better understood by their type and related characteristics.

Definitions

A few definitions:

Whole numbers: describes a set of numbers that does not contain any fractions or decimals. The set of whole numbers includes zero.

Example: 0, 1, 2, 3, 4, 189, 293 are all whole numbers.

Integers: describes whole numbers and their negative counterparts. (Zero does not have a negative counterpart here. Instead, zero is its own negative.)

Example: –1, –2, –3, –4, –5, 0, 1, 2, 3, 4, 5 are all integers.

–1, –2, –3, –4, –5 are considered negative integers, and 1, 2, 3, 4, 5 are considered positive integers.

Absolute value: describes the value of a number regardless of its sign. The symbol for absolute value is | |.

Example: The absolute value of 24 is 24 or |24| = 24.

The absolute value of –693 is 693 or |–693| = 693.

Even numbers: describes any number that can be divided by 2 evenly, meaning the answer has no decimal or remainder portion.

Example: 2, 4, 9082, –2, –16, –504 are all considered even numbers, because they can be divided by 2, without leaving a remainder or forming a decimal. It does not matter whether the number is positive or negative.

Odd numbers: describes any number that does not divide evenly by 2.

Example: 1, 21, 541, 3003, –9, –63, –1257 are all considered odd numbers, because they cannot be divided by 2 without a remainder or a decimal.

Prime numbers: describes a number that is only evenly divisible, resulting in no remainder or decimal, by 1 and itself.

Example: 2, 3, 7, 13, 113 are all considered prime numbers because each can only be evenly divided by 1 and itself.

Composite numbers: describes a positive integer that is formed by multiplying two smaller integers together. Composite numbers can be divided evenly by numbers other than 1 or itself.

Example: 9, 24, 66, 2348, 1,0002 are all considered composite numbers because they are the result of multiplying two smaller integers together. In particular, these are all divisible by 2.

Decimals: designated by a decimal point which indicates that what follows the point is a value that is less than 1 and is added to the integer number preceding the decimal point. The digit immediately following the decimal point is in the tenths place, the digit following the tenths place is in the hundredths place, and so on.

For example, the decimal number 1.735 has a value greater than 1 but less than 2. The 7 represents seven tenths of the unit 1 (0.7 or $\frac{7}{10}$); the 3 represents three hundredths of 1 (0.03 or $\frac{3}{100}$); and the 5 represents five thousandths of 1 (0.005 or $\frac{5}{1,000}$).

Real numbers: describes rational numbers and irrational numbers.

Rational numbers: describes any number that can be expressed as a fraction, with a non-zero denominator. Since any integer can be written with 1 in the denominator without changing its value, all integers are considered rational numbers. Every rational number has a decimal expression that terminates or repeats. That is, any rational number either will have a countable number of nonzero digits or will end with an ellipsis or a bar (3.6666... or $3.\bar{6}$) to depict repeating decimal digits. Some examples of rational numbers include 12, –3.54, $110.\overline{256}$, $\frac{-35}{10}$, and $4.\overline{7}$.

Irrational numbers: describes numbers that cannot be written as a finite decimal. Pi (π) is considered to be an irrational number because its decimal portion is unending or a non-repeating decimal. The most common irrational number is π, which has an endless and non-repeating decimal, but there are other well-known irrational numbers like *e* and $\sqrt{2}$.

Basic Addition and Subtraction

Addition

Addition is the combination of two numbers so their quantities are added together cumulatively. The sign for an addition operation is the + symbol. For example, 9 + 6 = 15. The 9 and 6 combine to achieve a cumulative value, called a **sum**.

Addition holds the **commutative property**, which means that the order of the numbers in an addition equation can be switched without altering the result. The formula for the commutative property is a + b = b + a. Let's look at a few examples to see how the commutative property works:

$$7 = 3 + 4 = 4 + 3 = 7$$

$$20 = 12 + 8 = 8 + 12 = 20$$

Addition also holds the **associative property**, which means that the grouping of numbers doesn't matter in an addition problem. In other words, the presence or absence of parentheses is irrelevant. The formula for the associative property is (a + b) + c = a + (b + c). Here are some examples of the associative property at work:

$$30 = (6 + 14) + 10 = 6 + (14 + 10) = 30$$

$$35 = 8 + (2 + 25) = (8 + 2) + 25 = 35$$

There are set columns for addition: ones, tens, hundreds, thousands, ten-thousands, hundred-thousands, millions, and so on. To add how many units there are total, each column needs to be combined, starting from the right, or the ones column.

THOUSANDS	HUNDREDS	TENS	ONES

Every 10 units in the ones column equals one in the tens column, and every 10 units in the tens column equals one in the hundreds column, and so on.

Example: The number 5,432 has 2 ones, 3 tens, 4 hundreds, and 5 thousands. The number 371 has 3 hundreds, 7 tens and 1 one. To combine, or add, these two numbers, simply add up how many units of each column exist. The best way to do this is by lining up the columns:

$$\begin{array}{r} 5\,4\,3\,2 \\ +\quad 3\,7\,1 \\ \hline \end{array}$$

The ones column adds 2 + 1 for a total (sum) of 3.

The tens column adds 3 + 7 for a total of 10; since 10 of that unit was collected, add 1 to the hundreds column to denote the total in the next column:

$$\begin{array}{r} 1\quad\; \\ 5\,4\,3\,2 \\ +\quad 3\,7\,1 \\ \hline 0\,3 \end{array}$$

When adding the hundreds column, this extra 1 needs to be combined, so it would be the sum of 4, 3, and 1.

$$4 + 3 + 1 = 8$$

The last, or thousands, column listed would be the sum of 5. Since there are no other numbers in this column, that is the final total.

The answer would look as follows:

$$\begin{array}{r} 5\,4\,3\,2 \\ +\quad 3\,7\,1 \\ \hline 5\,8\,0\,3 \end{array}$$

Example

Find the sum of 9,734 and 895.

Set up the problem:

$$\begin{array}{r} 9\,7\,3\,4 \\ +\quad 8\,9\,5 \\ \hline \end{array}$$

Total the columns:

$$\begin{array}{r} 9\ 7\ 3\ 4 \\ +\quad 8\ 9\ 5 \\ \hline 1\ 0\ 6\ 2\ 9 \end{array}$$

In this example, another column (ten-thousands) is added to the left of the thousands column, to denote a carryover of 10 units in the thousands column. The final sum is 10,629.

When adding using all negative integers, the total is negative. The integers are simply added together and the negative symbol is tacked on.

$$(-12) + (-435) = -447$$

Subtraction

Subtraction is taking away one number from another, so their quantities are reduced. The sign designating a subtraction operation is the – symbol, and the result is called the **difference**. For example, 9 - 6 = 3. The number *6* detracts from the number *9* to reach the difference *3*.

Unlike addition, subtraction follows neither the commutative nor associative properties. The order and grouping in subtraction impact the result.

$$15 = 22 - 7 \neq 7 - 22 = -15$$

$$3 = (10 - 5) - 2 \neq 10 - (5 - 2) = 7$$

When working through subtraction problems involving larger numbers, it's necessary to regroup the numbers. Let's work through a practice problem using regrouping:

$$\begin{array}{r} 3\ 2\ 5 \\ -\ 7\ 7 \\ \hline \end{array}$$

Here, it is clear that the ones and tens columns for 77 are greater than the ones and tens columns for 325. To subtract this number, borrow from the tens and hundreds columns. When borrowing from a column, subtracting 1 from the lender column will add 10 to the borrower column:

$$\begin{array}{rrr} 3\text{-}1 & 10\text{+}2\text{-}1 & 10\text{+}5 \\ - & 7 & 7 \\ \hline \end{array} = \begin{array}{rrr} 2 & 11 & 15 \\ - & 7 & 7 \\ \hline 2 & 4 & 8 \end{array}$$

After ensuring that each digit in the top row is greater than the digit in the corresponding bottom row, subtraction can proceed as normal, and the answer is found to be 248.

Addition and Subtraction with Negative Integers

When adding mixed-sign integers, determine which integer has the larger absolute value. Absolute value is the distance of a number from zero on the number line. Absolute value is indicated by these symbols: | |.

Take this equation for example:

$$12 + (-435)$$

The absolute value of each of the numbers is as follows:

$$|12| = 12$$

$$|-435| = 435$$

Since –435 is the larger integer, the final number will have its sign. In this case, that sign is negative. Now, subtract the smaller integer from the larger one. If this equation is worked out, it will look like this:

$$12 + (-435) = -423$$

Mathematically, the equation looks like the one above, but practically speaking it will be done it like this:

$$435 - 12 = 423$$

(then add the negative sign)

When subtracting with negative integers, every unmarked integer is assumed to have a positive sign. Subtracting an integer is the same as adding a negative integer.

Example:

–3 - 4
–3 + (–4)
–3 + (–4) = –7

Subtracting a negative integer is the same as adding a positive integer.

Example

–3 - (–4)
–3 + 4
–3 + 4 = 1

Multiplication of Whole Numbers

Multiplication involves adding together multiple copies of a number. It is indicated by an × symbol or a number immediately outside of a parenthesis. For example:

$$5(8 - 2)$$

The two numbers being multiplied together are called **factors**, and their result is called a **product**. For example, $9 \times 6 = 54$. This can be shown alternatively by expansion of either the 9 or the 6:

$$9 \times 6 = 9 + 9 + 9 + 9 + 9 + 9 = 54$$

$$9 \times 6 = 6 + 6 + 6 + 6 + 6 + 6 + 6 + 6 + 6 = 54$$

Like addition, multiplication holds the commutative and associative properties:

$$115 = 23 \times 5 = 5 \times 23 = 115$$

$$84 = 3 \times (7 \times 4) = (3 \times 7) \times 4 = 84$$

Multiplication also follows the **distributive property**, which allows the multiplication to be distributed through parentheses. The formula for distribution is $a \times (b + c) = ab + ac$. This is clear after the examples:

$$45 = 5 \times 9 = 5(3 + 6) = (5 \times 3) + (5 \times 6) = 15 + 30 = 45$$

$$20 = 4 \times 5 = 4(10 - 5) = (4 \times 10) - (4 \times 5) = 40 - 20 = 20$$

For larger-number multiplication, the way the numbers are lined up can make it easier to obtain the product. It is simplest to put the number with the most digits on top and the number with fewer digits on the bottom. If they have the same number of digits, select one for the top and one for the bottom. Line up the problem, and begin by multiplying the far-right column on the top and the far-right column on the bottom. If the answer to a column is more than 9, the ones place digit will be written below that column and the tens place digit will carry to the top of the next column to be added after those digits are multiplied. Write the answer below that column. Move to the next column to the left on the top, and multiply it by the same far-right column on the bottom. Keep moving to the left one column at a time on the top number until the end.

Example

Multiply 37×8

Line up the numbers, placing the one with the most digits on top.

$$\begin{array}{r} 3\ 7 \\ \times \quad 8 \\ \hline \end{array}$$

Multiply the far-right column on the top with the far-right column on the bottom (7 x 8). Write the answer, 56, as below: The ones value, 6, gets recorded, the tens value, 5, is carried.

$$\begin{array}{r} {}^{+5}\quad \\ 3\ 7 \\ \times \quad 8 \\ \hline 6 \end{array}$$

Move to the next column left on the top number and multiply with the far-right bottom (3 x 8). Remember to add any carry over after multiplying: 3 x 8 = 24, 24 + 5 = 29. Since there are no more digits on top, write the entire number below.

$$\begin{array}{r} {}^{+5} \\ 3\ 7 \\ \text{X}\quad 8 \\ \hline 2\ 9\ 6 \end{array}$$

The solution is 296.

If there is more than one column to the bottom number, move to the row below the first strand of answers, mark a zero in the far-right column, and then begin the multiplication process again with the far-right column on top and the second column from the right on the bottom. For each digit in the bottom number, there will be a row of answers, each padded with the respective number of zeros on the right. Finally, add up all of the answer rows for one total number.

Example: Multiply 512×36.

Line up the numbers (the one with the most digits on top) to multiply.

Begin with the right column on top and the right column on bottom (2×6).

$$\begin{array}{r} 5\ 1\ 2 \\ \text{X}\quad 3\ 6 \\ \hline \end{array}$$

Move one column left on top and multiply by the far-right column on the bottom (1×6). Add the carry over after multiplying: $1 \times 6 = 6, 6 + 1 = 7$.

$$\begin{array}{r} {}^{+1} \\ 5\ 1\ 2 \\ \text{x}\quad 3\ 6 \\ \hline 7\ 2 \end{array}$$

Move one column left on top and multiply by the far-right column on the bottom (5×6). Since this is the last digit on top, write the whole answer below.

$$\begin{array}{r} 5\ 1\ 2 \\ \text{X}\quad 3\ 6 \\ \hline 3\ 0\ 7\ 2 \end{array}$$

Now move on to the second column on the bottom number. Starting on the far-right column on the top, repeat this pattern for the next number left on the bottom (2×3). Write the answers below the first line of answers; remember to begin with a zero placeholder on the far right.

$$\begin{array}{r} 5\ 1\ 2 \\ \text{X} \quad 3\ 6 \\ \hline 3\ 0\ 7\ 2 \\ 6\ 0 \end{array}$$

Continue the pattern (1×3).

$$\begin{array}{r} 5\ 1\ 2 \\ \text{X} \quad 3\ 6 \\ \hline 3\ 0\ 7\ 2 \\ 3\ 6\ 0 \end{array}$$

Since this is the last digit on top, write the whole answer below.

$$\begin{array}{r} 5\ 1\ 2 \\ \text{x} \quad 3\ 6 \\ \hline 3\ 0\ 7\ 2 \\ 1\ 5\ 3\ 6\ 0 \end{array}$$

Now add the answer rows together. Pay attention to ensure they are aligned correctly.

$$\begin{array}{r} 5\ 1\ 2 \\ \text{x} \quad 3\ 6 \\ \hline 3\ 0\ 7\ 2 \\ 1\ 5\ 3\ 6\ 0 \\ \hline 1\ 8\ 4\ 3\ 2 \end{array}$$

The solution is 18,432.

Division of Whole Numbers

Division and multiplication are inverses of each other in the same way that addition and subtraction are opposites. The signs designating a division operation are the ÷ and / symbols. In division, the second number divides into the first.

The number before the division sign is called the **dividend** or, if expressed as a fraction, the **numerator**. For example, in $a \div b$, a is the dividend, while in $\frac{a}{b}$, a is the numerator.

The number after the division sign is called the **divisor** or, if expressed as a fraction, the **denominator.** For example, in $a \div b$, b is the divisor, while in $\frac{a}{b}$, b is the denominator.

Like subtraction, division doesn't follow the commutative property, as it matters which number comes before the division sign, and division doesn't follow the associative or distributive properties for the same reason. For example:

$$\frac{3}{2} = 9 \div 6 \neq 6 \div 9 = \frac{2}{3}$$

$$2 = 10 \div 5 = (30 \div 3) \div 5 \neq 30 \div (3 \div 5) = 30 \div \frac{3}{5} = 50$$

$$25 = 20 + 5 = (40 \div 2) + (40 \div 8) \neq 40 \div (2 + 8) = 40 \div 10 = 4$$

The answer to a division problem is called the **quotient.** If a divisor doesn't divide into a dividend evenly, whatever is left over is termed the **remainder**. The remainder can be further divided out into decimal form by using long division; however, this doesn't always give a quotient with a finite number of decimal places, so the remainder can also be expressed as a fraction over the original divisor.

Example

Solve $1050/42$ or $1050 \div 42$.

Set up the problem with the denominator being divided into the numerator.

$$42\overline{)1050}$$

Check for divisibility into the first unit of the numerator, 1.

42 cannot go into 1, so add on the next unit in the denominator, 0.

42 cannot go into 10, so add on the next unit in the denominator, 5.

42 can be divided into 105 two times. Write the 2 over the 5 in 105 and multiply 42 x 2. Write the 84 under 105 for subtraction and note the remainder, 21 is less than 42.

$$\begin{array}{r} 2 \\ 42\overline{)1050} \\ -84 \\ \hline 21 \end{array}$$

Drop the next digit in the numerator down to the remainder (making 21 into 210) to create a number 42 can divide into. 42 divides into 210 five times. Write the 5 over the 0 and multiply 42×5.

$$\begin{array}{r} 25 \\ 42\overline{)1050} \\ -84 \\ \hline 210 \end{array}$$

Write the 210 under 210 for subtraction. The remainder is 0.

$$\begin{array}{r} 25 \\ 42\overline{)1050} \\ -84 \\ \hline 210 \\ -210 \\ \hline 0 \end{array}$$

The solution is 25.

Example
Divide $375/4$ or $375 \div 4$.

Set up the problem.

$$4\overline{)375}$$

4 cannot divide into 3, so add the next unit from the numerator, 7. 4 divides into 37 nine times, so write the 9 above the 7. Multiply $4 \times 9 = 36$. Write the 36 under the 37 for subtraction. The remainder is 1 (1 is less than 4).

$$\begin{array}{r} 9 \\ 4\overline{)375} \\ -36 \\ \hline 1 \end{array}$$

Drop the next digit in the numerator, 5, making the remainder 15. 4 divides into 15, three times, so write the 3 above the 5. Multiply 4×3. Write the 12 under the 15 for subtraction, remainder is 3 (3 is less than 4).

$$\begin{array}{r} 93 \\ 4\overline{)375} \\ -36 \\ \hline 15 \\ -12 \\ \hline 3 \end{array}$$

The solution is 93 remainder 3 or 93 ¾ (the remainder can be written over the original denominator).

Distance Between Numbers on a Number Line

Aside from zero, numbers can be either positive or negative. The sign for a positive number is the plus sign or the + symbol, while the sign for a negative number is the minus sign or the – symbol. If a number has no designation, then it's assumed to be positive.

Both positive and negative numbers are valued according to their distance from zero. Both +3 and –3 can be considered using the following number line:

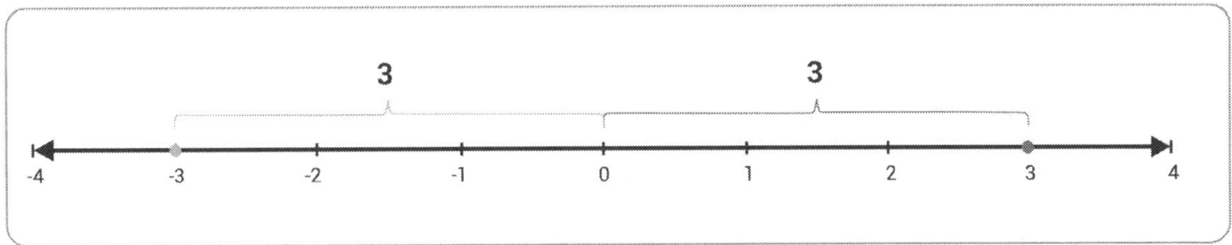

Both 3 and –3 are three spaces from zero. The distance from zero is called its **absolute value**. Thus, both –3 and 3 have an absolute value of 3 since they're both three spaces away from zero.

An absolute number is written by placing | | around the number. So, $|3|$ and $|-3|$ both equal 3, as that's their common absolute value.

Implications for Addition and Subtraction

For addition, if all numbers are either positive or negative, they are simply added together. For example, $4 + 4 = 8$ and $-4 + -4 = -8$. However, things get tricky when some of the numbers are negative, and some are positive.

For example, with $6 + (-4)$, the first step is to take the absolute values of the numbers, which are 6 and 4. Second, the smaller value is subtracted from the larger. The equation becomes $6 - 4 = 2$. Third, the sign of the original larger number is placed on the sum. Here, 6 is the larger number, and it's positive, so the sum is 2.

Here's an example where the negative number has a larger absolute value: $(-6) + 4$. The first two steps are the same as the example above. However, on the third step, the negative sign must be placed on the sum, because the absolute value of (–6) is greater than 4. Thus, $-6 + 4 = -2$.

The absolute value of numbers implies that subtraction can be thought of as flipping the sign of the number following the subtraction sign and simply adding the two numbers. This means that subtracting a negative number will, in fact, be adding the positive absolute value of the negative number.

Here are some examples:

$$-6 - 4 = -6 + -4 = -10$$

$$3 - -6 = 3 + 6 = 9$$

$$-3 - 2 = -3 + -2 = -5$$

Implications for Multiplication and Division

For multiplication and division, if both numbers are positive, then the product or quotient is always positive. If both numbers are negative, then the product or quotient is also positive. However, if the numbers have opposite signs, the product or quotient is always negative.

Simply put, the product in multiplication and quotient in division is always positive, unless the numbers have opposing signs, in which case it's negative. Here are some examples:

$$(-6) \times (-5) = 30$$

$$(-50) \div 10 = -5$$

$$8 \times |-7| = 56$$

$$(-48) \div (-6) = 8$$

If there are more than two numbers in a multiplication or division problem, then whether the product or quotient is positive or negative depends on the number of negative numbers in the problem. If there is an odd number of negatives, then the product or quotient is negative. If there is an even number of negative numbers, then the result is positive.

Here are some examples:

$$(-6) \times 5 \times (-2) \times (-4) = -240$$

$$(-6) \times 5 \times 2 \times (-4) = 240$$

Multiples and Factors

Multiples of a given number are found by taking that number and multiplying it by any other whole number. For example, 3 is a factor of 6, 9, and 12. Therefore, 6, 9, and 12 are multiples of 3. The multiples of any number are an infinite list. For example, the multiples of 5 are 5, 10, 15, 20, and so on. This list continues without end. A list of multiples is used in finding the **least common multiple**, or LCM, for fractions when a common denominator is needed. The denominators are written down and their multiples listed until a common number is found in both lists. This common number is the LCM.

The **factors** of a number are all integers that can be multiplied by another integer to produce the given number. For example, 2 is multiplied by 3 to produce 6. Therefore, 2 and 3 are both factors of 6. Similarly, $1 \times 6 = 6$ and $2 \times 3 = 6$, so 1, 2, 3, and 6 are all factors of 6. Another way to explain a factor is to say that a given number divides evenly by each of its factors to produce an integer. For example, 6 does not divide evenly by 5. Therefore, 5 is not a factor of 6.

Prime factorization breaks down each factor of a whole number until only prime numbers remain. All composite numbers can be factored into prime numbers. For example, the prime factors of 12 are 2, 2, and:

$$3\ (2 \times 2 \times 3 = 12)$$

To produce the prime factors of a number, the number is factored, and any composite numbers are continuously factored until the result is the product of prime factors only. A **factor tree**, such as the one below, is helpful when exploring this concept.

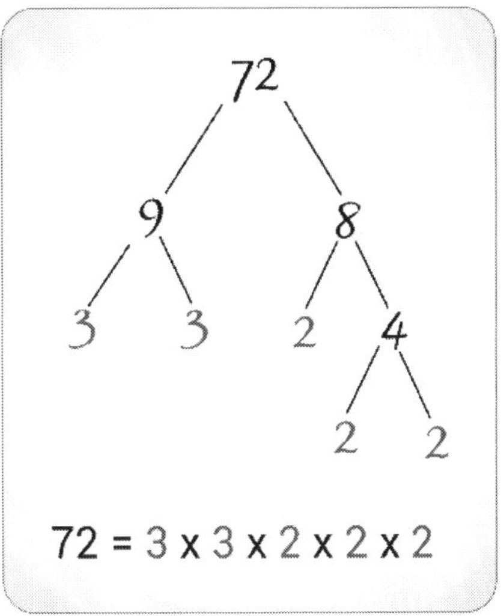

Properties of Operations

Properties of operations exist to make calculations easier and solve problems for missing values. The following table summarizes commonly used properties of real numbers.

Property	Addition	Multiplication
Commutative	$a + b = b + a$	$a \times b = b \times a$
Associative	$(a + b) + c = a + (b + c)$	$(a \times b) \times c = a \times (b \times c)$
Identity	$a + 0 = a;\ 0 + a = a$	$a \times 1 = a;\ 1 \times a = a$
Inverse	$a + (-a) = 0$	$a \times \frac{1}{a} = 1;\ a \neq 0$
Distributive	$a(b + c) = ab + ac$	

The **commutative property of addition** states that the order in which numbers are added does not change the sum. Similarly, the **commutative property of multiplication** states that the order in which numbers are multiplied does not change the product. The **associative property of addition** and **multiplication** state that the grouping of numbers being added or multiplied does not change the sum or product, respectively. The commutative and associative properties are useful for performing calculations. For example, $(47 + 25) + 3$ is equivalent to $(47 + 3) + 25$, which is easier to calculate.

The **identity property of addition** states that adding zero to any number does not change its value. The **identity property of multiplication** states that multiplying a number by one does not change its value. The **inverse property of addition** states that the sum of a number and its opposite equals zero. Opposites are numbers that are the same with different signs (ex. 5 and –5; $-\frac{1}{2}$ and $\frac{1}{2}$). The **inverse property of multiplication** states that the product of a number (other than zero) and its reciprocal equals one. **Reciprocal numbers** have numerators and denominators that are inverted (ex. $\frac{2}{5}$ and $\frac{5}{2}$).

Inverse properties are useful for canceling quantities to find missing values (see algebra content). For example, $a + 7 = 12$ is solved by adding the inverse of $7(-7)$ to both sides in order to isolate a.

The **distributive property states** that multiplying a sum (or difference) by a number produces the same result as multiplying each value in the sum (or difference) by the number and adding (or subtracting) the products. Consider the following scenario: You are buying three tickets for a baseball game. Each ticket costs $18. You are also charged a fee of $2 per ticket for purchasing the tickets online. The cost is calculated:

$$3 \times 18 + 3 \times 2$$

Using the distributive property, the cost can also be calculated $3(18 + 2)$.

Prefixes

Moving the decimal place to the left or to the right illustrates multiplying or dividing by factors of 10. The metric system of units for measurement utilizes factors of 10 as displayed in the following table:

kilo	1,000 units
hecto	100 units
deca	10 units
base unit	
deci	0.1 units
centi	0.01 units
milli	0.001 units

It is important to have the ability to quickly manipulate by 10 according to prefixes for units.

Example: How many milliliters are in 5 liters of saline solution?

There are 1,000 milliliters for every 1 liter. If we have 5 liters, it would be $5 \times 1{,}000 = 5000$ mL

You may also count the zeros and which side of the decimal place they are on: 1,000 has three zeroes to the left of the decimal, so insert three zeroes between the 5 and the decimal, or move the decimal place over three places to the right, for your answer of 5000 mL.

Example

How many kilograms are in 4.8 grams?

There is 1 gram for every 0.001 kilograms. Since there is one-thousandth of a kilogram for each gram, that means divide by 1,000, or move the decimal to the left by 3 places – 1 place for each 0. So, the result would be 0.0048 kg.

For quick conversions, move the decimal place the set number of spaces left or right to match the column/slot, as depicted below.

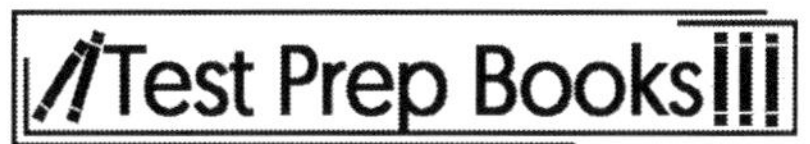

To convert from one prefix to another to the left or right of the base unit (follow the arrow to the left or right), move the decimal place the number of columns/slots as counted.

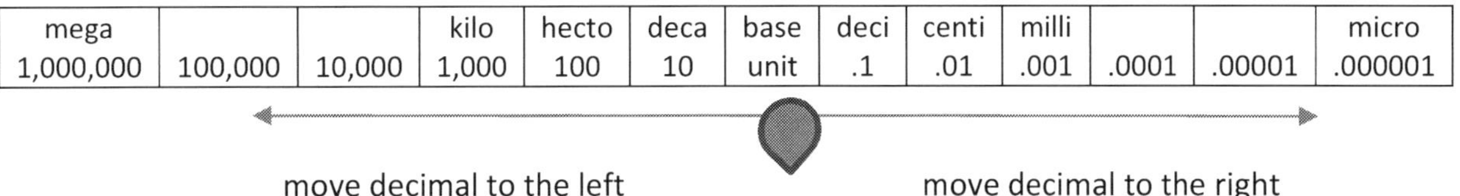

mega			kilo	hecto	deca	base	deci	centi	milli			micro
1,000,000	100,000	10,000	1,000	100	10	unit	.1	.01	.001	.0001	.00001	.000001

Example

How many centiliters are in 4.7 kiloliters?

To convert a number with a unit prefixed as kilo into a unit prefixed as centi, move across five columns to the right, meaning move the decimal place five places to the right.

4.7 kL = 470,000 cL

Example

How many liters are in 30 microliters?

Start with the unit marked micro and count the columns moving to the left until you reach the base unit for liters. Be sure to count the blank columns, as they are important placeholders. There are six columns from micro to the base unit moving to the left, so move the decimal place six places to the left.

30 mL = 0.000030 L

Decimals

Decimals mark the separation between the whole portion and the fractional (or decimal) portion of a number. For example, 3.15 has 3 in the whole portion and 15 in the fractional or decimal portion. A number such as 645 is all whole, but there is still a decimal place. The decimal place in 645 is to the right of the 5, but usually not written, since there is no fractional or decimal portion to this number. The same number can be written as 645.0 or 645.00 or 645.000, etc. The position of the decimal place can change the entire value of a number, and impact a calculation. In the United States, the decimal place is used when representing money. You'll often be asked to round to a certain decimal place. Here is a review of some basic decimal **place value** names:

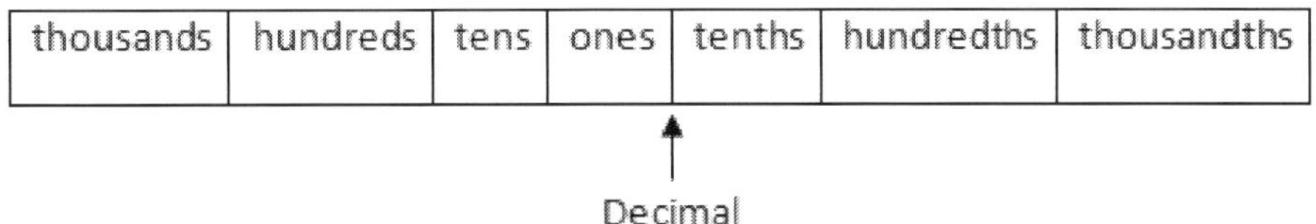

thousands	hundreds	tens	ones	tenths	hundredths	thousandths

The number 12,302.2 would be read as "twelve thousand, three hundred two and two-tenths."

In the United States, a period denotes the decimal place; however, some countries use a comma. The comma is used in the United States to separate thousands, millions, and so on.

To round to the nearest whole number (eliminating the decimal portion), the example would become 12,302. For rounding, go to the number that is one place to the right of what you are rounding to. If the number is 0 through 4, there will be no change. For numbers 5 through 9, round up to the next whole number.

Example
Round 6,423.7 to the ones place.

Since the tenths place is the position to the right of the ones place, we use that number to determine if we round up or not. In this case, the 3 is in the ones place and the 7 is in the tenths place. (6,423.7)

The 7 in the tenths place means we round the 3 up, so the final number will be 6,424.0

Example
Round 542.88 to the nearest tens

Since the ones place is the position to the right of the tens, we use that number to determine if we round up or not. In this case, the 4 is in the tens place and the 2 is in the ones place (542.88).

The 2 in the ones place means we do not round the 4 up, so the final number will be 540.00

Note: Everything to the right of the rounded position goes to 0 as a placeholder.

Example: Say you wanted to post an advertisement to sell a used vehicle for $2000.00. However, when typing the price, you accidentally moved the decimal over one place to the left. Now the asking price appears as $200.00. This difference of a factor of 10 is dramatic. As numbers get bigger or smaller, the impact of this mistake becomes more pronounced. If you were looking to sell a condo for $1,000,000.00, but made an error and moved the decimal place to the left one position, the price posts at $100,000.00. A mistake of a factor of 10 cost $900,000.00.

In dividing by 10, you move the decimal one position to the left, making a smaller number than the original. If multiplying by 10, move the decimal one position to the right, making a larger number than the original.

Example
Divide 100 by 10 or 100 ÷ 10.

Move the decimal one place to the left, so the result is a smaller number than the original.

$$100 \div 10 = 10$$

Example
Divide 1.0 by 10 or 1.0 ÷ 10.

Move the decimal one place to the left, so the result is a smaller number than the original.

$$1.0 \div 10 = 0.1$$

Example
Multiply 100 by 10 or 100 x 10.

Move the decimal one place to the right, so the result is a larger number than the original.

$$100 \times 10 = 1{,}000$$

Example
Multiply 0.1 by 10 or 0.1 x 10.

Move the decimal one place to the right, so the result is a larger number than the original.

$$0.1 \times 10 = 1.0$$

Decimal Addition

Addition with decimals is done the same way as regular addition. All numbers could have decimals, but are often removed if the numbers to the right of the decimal are zeros. Line up numbers at the decimal place.

Example: Add $345.89 + 23.54$

Line the numbers up at the decimal place and add.

$$\begin{array}{r} 345.89 \\ +\ \ 23.54 \\ \hline 369.43 \end{array}$$

Decimal Subtraction

Subtraction with decimals is done the same way as regular subtraction.

Example: Subtract $345.89 - 23.54$

Line the numbers up at the decimal place and subtract.

$$\begin{array}{r} 345.89 \\ -\ \ 23.54 \\ \hline 322.35 \end{array}$$

Decimal Multiplication

The simplest way to handle multiplication with decimals is to calculate the multiplication problem pretending the decimals are not there, then count how many decimal places there are in the original problem. Use that total to place the decimal the same number of places over, counting from right to left.

Example: Multiply 42.33×3.3

Line the numbers up and multiply, pretending there are no decimals.

$$\begin{array}{r} 4\,2\,3\,3 \\ \times \quad 3\,3 \\ \hline 1\,2\,6\,9\,9 \\ 1\,2\,6\,9\,9\,0 \\ \hline 1\,3\,9\,6\,8\,9 \end{array}$$

Now look at the original problem and count how many decimal places were removed. Two decimal places were removed from 42.33 to get 4233, and one decimal place from 3.3 to get 33. Removed were $2 + 1 = 3$ decimal places. Place the decimal three places from the right of the number 139689. The answer is 139.689.

Another way to think of this is that when you move the decimal in the original numbers, it is like multiplying by 10. To put the decimals back, you need to divide the number by 10 the same amount of times you multiplied. It would still be three times for the above solution.

Example: Multiply 0.03×1.22

Line the numbers up and multiply, pretending there are no decimals. The zeroes in front of the 3 are unnecessary, so take them out for now.

$$\begin{array}{r} 1\,2\,2 \\ \times \quad 3 \\ \hline 3\,6\,6 \end{array}$$

Look at the original problem and count how many decimals places were removed, or how many times each number was multiplied by 10. The 1.22 moved two places (or multiplied by 10 twice), as did 0.03. That is $2 + 2 = 4$ decimal places removed. Count that number, from right to left of the number 366, and place the decimal. The result is 0.0366.

Decimal Division

Division with decimals is simplest when you eliminate some of the decimal places. Since you divide the bottom number of a fraction into the top, or divide the denominator into the numerator, the bottom number dictates the movement of the decimals. The goal is to remove the decimals from the denominator and mirror that movement in the numerator. You do not need the numerator to be decimal free, however. Divide as you would normally.

Example

Divide $4.21/0.2$ or $4.21 \div 0.2$

Move the decimal over one place to the right in the denominator, making 0.2 simply 2. Move the decimal in the numerator, 4.21, over the same amount, so it is now 42.1.

$$0.2\overline{)4.21}$$

Becomes

$$2\overline{)42.1}$$

Divide.

$$\begin{array}{r} 21.05 \\ 2\overline{)42.10} \end{array}$$

The answer is 21.05 with the correct decimal placement. In decimal division, move the decimal the same amount for both numerator and denominator. There is no need to adjust anything after the problem is completed.

Fractions

A fraction is an equation that represents a part of a whole but can also be used to present ratios or division problems. An example of a fraction is $\frac{x}{y}$. In this example, x is called the **numerator**, while y is the **denominator**. The numerator represents the number of parts, and the denominator is the total number of parts. They are separated by a line or slash, known as a **fraction bar**. In simple fractions, the numerator and denominator can be nearly any integer. However, the denominator of a fraction can never be zero because dividing by zero is a function that is undefined.

Imagine that an apple pie has been baked for a holiday party, and the full pie has eight slices. After the party, there are five slices left. How could the amount of the pie that remains be expressed as a fraction? The numerator is 5 since there are 5 pieces left, and the denominator is 8 since there were eight total slices in the whole pie. Thus, expressed as a fraction, the leftover pie totals $\frac{5}{8}$ of the original amount.

Fractions come in three different varieties: proper fractions, improper fractions, and mixed numbers. **Proper fractions** have a numerator less than the denominator, such as $\frac{3}{8}$, but **improper fractions** have a numerator greater than the denominator, such as $\frac{15}{8}$. **Mixed numbers** combine a whole number with a proper fraction, such as $3\frac{1}{2}$. Any mixed number can be written as an improper fraction by multiplying the integer by the denominator, adding the product to the value of the numerator, and dividing the sum by the original denominator. For example:

$$3\frac{1}{2} = \frac{3 \times 2 + 1}{2} = \frac{7}{2}$$

Whole numbers can also be converted into fractions by placing the whole number as the numerator and making the denominator 1. For example, $3 = \frac{3}{1}$.

One of the most fundamental concepts of fractions is their ability to be manipulated by multiplication or division. This is possible since $\frac{n}{n} = 1$ for any non-zero integer. As a result, multiplying or dividing by $\frac{n}{n}$ will not alter the original fraction since any number multiplied or divided by 1 doesn't change the value of that number. Fractions of the same value are known as equivalent fractions. For example, $\frac{2}{4}, \frac{4}{8}, \frac{50}{100}$, and $\frac{75}{150}$ are equivalent, as they all equal $\frac{1}{2}$.

Although many equivalent fractions exist, they are easier to compare and interpret when reduced or simplified. The numerator and denominator of a simple fraction will have no factors in common other than 1. When reducing or simplifying fractions, divide the numerator and denominator by the **greatest common factor**. A simple strategy is to divide the numerator and denominator by low numbers, like 2, 3, or 5 until arriving at a simple fraction, but the same thing could be achieved by determining the greatest common factor for both the numerator and denominator and dividing each by it. Using the first method is preferable when both the numerator and denominator are even, end in 5, or are obviously a multiple of another number. However, if no numbers seem to work, it will be necessary to factor the numerator and denominator to find the GCF. For example:

1) Simplify the fraction $\frac{6}{8}$:

Dividing the numerator and denominator by 2 results in $\frac{3}{4}$, which is a simple fraction.

2) Simplify the fraction $\frac{12}{36}$:

Dividing the numerator and denominator by 2 leaves $\frac{6}{18}$. This isn't a simple fraction, as both the numerator and denominator have factors in common. Dividing each by 3 results in $\frac{2}{6}$, but this can be further simplified by dividing by 2 to get $\frac{1}{3}$. This is the simplest fraction, as the numerator is 1. In cases like this, multiple division operations can be avoided by determining the greatest common factor between the numerator and denominator.

3) Simplify the fraction $\frac{18}{54}$ by dividing by the greatest common factor:

First, determine the factors for the numerator and denominator. The factors of 18 are 1, 2, 3, 6, 9, and 18. The factors of 54 are 1, 2, 3, 6, 9, 18, 27, and 54. Thus, the greatest common factor is 18. Dividing both the numerator and denominator by 18 leaves $\frac{1}{3}$, which is the simplest fraction. This method takes slightly more work, but it definitively arrives at the simplest fraction.

Operations with Fractions

Multiplication of Fractions

Of the four basic operations that can be performed on fractions, the one that involves the least amount of work is multiplication. To multiply two fractions, simply multiply the numerators together, multiply the denominators together, and place the products of each as a fraction. Whole numbers and mixed numbers can also be expressed as a fraction, as described above, to multiply with a fraction. Here are a few examples:

$$1)\ \frac{2}{5} \times \frac{3}{4} = \frac{6}{20} = \frac{3}{10}$$

$$2)\ \frac{4}{9} \times \frac{7}{11} = \frac{28}{99}$$

Division of Fractions

Dividing fractions is similar to multiplication with one key difference. To divide fractions, flip the numerator and denominator of the second fraction, and then proceed as if it were a multiplication problem:

$$1)\,\frac{7}{8} \div \frac{4}{5} = \frac{7}{8} \times \frac{5}{4} = \frac{35}{32}$$

$$2)\,\frac{5}{9} \div \frac{1}{3} = \frac{5}{9} \times \frac{3}{1} = \frac{15}{9} = \frac{5}{3}$$

Addition and Subtraction of Fractions

Addition and subtraction require more steps than multiplication and division, as these operations require the fractions to have the same denominator, also called a **common denominator**. It is always possible to find a common denominator by multiplying the denominators. However, when the denominators are large numbers, this method is unwieldy, especially if the answer must be provided in its simplest form. Thus, it's beneficial to find the **least common denominator** of the fractions—the least common denominator is incidentally also the least common multiple.

Once equivalent fractions have been found with common denominators, simply add or subtract the numerators to arrive at the answer:

$$1)\,\frac{1}{2} + \frac{3}{4} = \frac{2}{4} + \frac{3}{4} = \frac{5}{4}$$

$$2)\,\frac{3}{12} + \frac{11}{20} = \frac{15}{60} + \frac{33}{60} = \frac{48}{60} = \frac{4}{5}$$

$$3)\,\frac{7}{9} - \frac{4}{15} = \frac{35}{45} - \frac{12}{45} = \frac{23}{45}$$

$$4)\,\frac{5}{6} - \frac{7}{18} = \frac{15}{18} - \frac{7}{18} = \frac{8}{18} = \frac{4}{9}$$

Changing Fractions to Decimals

To change a fraction into a decimal, divide the denominator into the numerator until there are no remainders. There may be repeating decimals, so rounding is often acceptable. A straight line above the repeating portion denotes that the decimal repeats.

Example

Express 4/5 as a decimal.

Set up the division problem.

$$5\overline{)4}$$

5 does not go into 4, so place the decimal and add a zero.

$$5\overline{)4.0}$$

5 goes into 40 eight times. There is no remainder.

$$\begin{array}{r} 0.8 \\ 5\overline{)4.0} \\ -\underline{4.0} \\ 0 \end{array}$$

The solution is 0.8.

Example

Express 33 1/3 as a decimal.

Since the whole portion of the number is known, set it aside to calculate the decimal from the fraction portion.

Set up the division problem.

$$3\overline{)1}$$

3 does not go into 1, so place the decimal and add zeros. 3 goes into 10 three times.

$$\begin{array}{r} 0.3 \\ 3\overline{)1.0} \end{array}$$

This will repeat with a remainder of 1.

$$\begin{array}{r} 0.333 \\ 3\overline{)1.000} \\ -\underline{9} \\ 10 \\ -\underline{9} \\ 10 \end{array}$$

So, we will place a line over the 3 to denote the repetition. The solution is written $33.\overline{3}$.

Changing Decimals to Fractions

To change decimals to fractions, place the decimal portion of the number, the numerator, over the respective place value, the denominator, then reduce, if possible.

Example
Express 0.25 as a fraction.

This is read as twenty-five hundredths, so put 25 over 100. Then reduce to find the solution.

$$\frac{25}{100} = \frac{1}{4}$$

Example
Express 0.455 as a fraction

This is read as four hundred fifty-five thousandths, so put 455 over 1,000. Then reduce to find the solution.

$$\frac{455}{1{,}000} = \frac{91}{200}$$

There are two types of problems that commonly involve percentages. The first is to calculate some percentage of a given quantity, where you convert the percentage to a decimal, and multiply the quantity by that decimal. Secondly, you are given a quantity and told it is a fixed percent of an unknown quantity. In this case, convert to a decimal, then divide the given quantity by that decimal.

Example
What is 30% of 760?

Convert the percent into a useable number. "Of" means to multiply.

$$30\% = 0.30$$

Set up the problem based on the givens, and solve.

$$0.30 \times 760 = 228$$

Example
8.4 is 20% of what number?

Convert the percent into a useable number.

$$20\% = 0.20$$

The given number is a percent of the answer needed, so divide the given number by this decimal rather than multiplying it.

$$\frac{8.4}{0.20} = 42$$

Fractions and Decimals in Order

A rational number is any number that can be written as a fraction or ratio. Within the set of rational numbers, several subsets exist that are referenced throughout the mathematics topics. Counting numbers are the first numbers learned as a child. Counting numbers consist of 1,2,3,4, and so on. Whole numbers include all counting numbers and zero (0,1,2,3,4,...). Integers include counting numbers, their opposites, and zero (...,–3,–2,–1,0,1,2,3,...). Rational numbers are inclusive of integers, fractions, and decimals that terminate, or end (1.7, 0.04213) or repeat ($0.136\overline{5}$).

Placing numbers in an order in which they are listed from smallest to largest is known as **ordering**. Ordering numbers properly can help in the comparison of different quantities of items.

When comparing two numbers to determine if they are equal or if one is greater than the other, it is best to look at the digit furthest to the left of the decimal place (or the first value of the decomposed numbers). If this first digit of each number being compared is equal in place value, then move one digit to the right to conduct a similar comparison. Continue this process until it can be determined that both numbers are equal or a difference is found, showing that one number is greater than the other. If a number is greater than the other number it is being compared to, a symbol such as > (greater than) or < (less than) can be utilized to show this comparison. It is important to remember that the "open mouth" of the symbol should be nearest the larger number.

For example:

1,023,100 compared to 1,023,000

First, compare the digit farthest to the left. Both are decomposed to 1,000,000, so this place is equal.

Next, move one place to right on both numbers being compared. This number is zero for both numbers, so move on to the next number to the right. The first number decomposes to 20,000, while the second decomposes to 20,000. These numbers are also equal, so move one more place to the right. The first number decomposes to 3,000, as does the second number, so they are equal again. Moving one place to the right, the first number decomposes to 100, while the second number is zero. Since 100 is greater than zero, the first number is greater than the second. This is expressed using the greater than symbol:

1,023,100 > 1,023,000 because 1,023,100 is greater than 1,023,000 (Note that the "open mouth" of the symbol is nearest to 1,023,100).

Notice the > symbol in the above comparison. When values are the same, the equals sign (=) is used. However, when values are unequal, or an **inequality** exists, the relationship is denoted by various inequality symbols. These symbols describe in what way the values are unequal. A value could be greater than ($>$); less than ($<$); greater than or equal to ($\geq$); or less than or equal to ($\leq$) another value. The statement "five times a number added to forty is more than sixty-five" can be expressed as $5x + 40 > 65$. Common words and phrases that express inequalities are:

Symbol	Phrase
$<$	is under, is below, smaller than, beneath
$>$	is above, is over, bigger than, exceeds
$\leq$	no more than, at most, maximum
$\geq$	no less than, at least, minimum

Another way to compare whole numbers with many digits is to use place value. In each number to be compared, it is necessary to find the highest place value in which the numbers differ and to compare the value within that place value. For example, $4{,}523{,}345 < 4{,}532{,}456$ because of the values in the ten thousands place.

Comparing and Ordering Decimals

To compare decimals and order them by their value, utilize a method similar to that of ordering large numbers.

The main difference is where the comparison will start. Assuming that any numbers to left of the decimal point are equal, the next numbers to be compared are those immediately to the right of the decimal point. If those are equal, then move on to compare the values in the next decimal place to the right.

For example:

Which number is greater, 12.35 or 12.38?

Check that the values to the left of the decimal point are equal:

$$12 = 12$$

Next, compare the values of the decimal place to the right of the decimal:

$$12.3 = 12.3$$

Those are also equal in value.

Finally, compare the value of the numbers in the next decimal place to the right on both numbers:

$$12.3\mathbf{5} \text{ and } 12.3\mathbf{8}$$

Here the 5 is less than the 8, so the final way to express this inequality is:

$$12.35 < 12.38$$

Comparing decimals is regularly exemplified with money because the "cents" portion of money ends in the hundredths place. When paying for gasoline or meals in restaurants, and even in bank accounts, if enough errors are made when calculating numbers to the hundredths place, they can add up to dollars and larger amounts of money over time.

Number lines can also be used to compare decimals. Tick marks can be placed within two whole numbers on the number line that represent tenths, hundredths, etc. Each number being compared can then be plotted. The value farthest to the right on the number line is the largest.

Comparing Fractions

To compare fractions with either the same **numerator** (top number) or same **denominator** (bottom number), it is easiest to visualize the fractions with a model.

For example, which is larger, $\frac{1}{3}$ or $\frac{1}{4}$? Both numbers have the same numerator, but a different denominator. In order to demonstrate the difference, shade the amounts on a pie chart split into the number of pieces represented by the denominator.

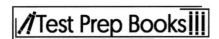

The first pie chart represents $\frac{1}{3}$, a larger shaded portion, and is therefore a larger fraction than the second pie chart representing $\frac{1}{4}$.

If two fractions have the same denominator (or are split into the same number of pieces), the fraction with the larger numerator is the larger fraction, as seen below in the comparison of $\frac{1}{3}$ and $\frac{2}{3}$:

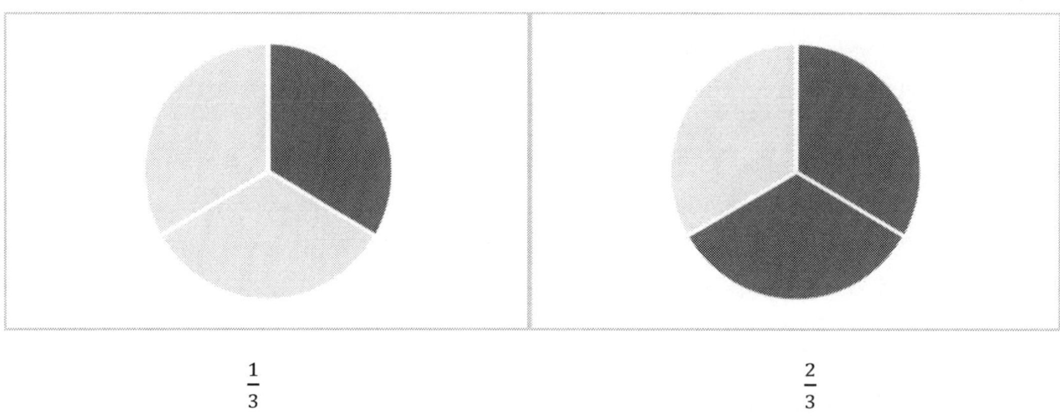

$\frac{1}{3}$ $\frac{2}{3}$

A **unit fraction** is one in which the numerator is 1 ($\frac{1}{2}, \frac{1}{3}, \frac{1}{8}, \frac{1}{20}$, etc.). The denominator indicates the number of equal pieces that the whole is divided into. The greater the number of pieces, the smaller each piece will be. Therefore, the greater the denominator of a unit fraction, the smaller it is in value. Unit fractions can also be compared by converting them to decimals. For example, $\frac{1}{2}$ = 0.5, $\frac{1}{3}$ = $0.\overline{3}$, $\frac{1}{8}$ = 0.125, $\frac{1}{20} = 0.05$, etc.

Comparing two fractions with different denominators can be difficult if attempting to guess at how much each represents. Using a number line, blocks, or just finding a common denominator with which to compare the two fractions makes this task easier.

For example, compare the fractions $\frac{3}{4}$ and $\frac{5}{8}$.

The number line method of comparison involves splitting one number line evenly into 4 sections, and the second number line evenly into 8 sections total, as follows:

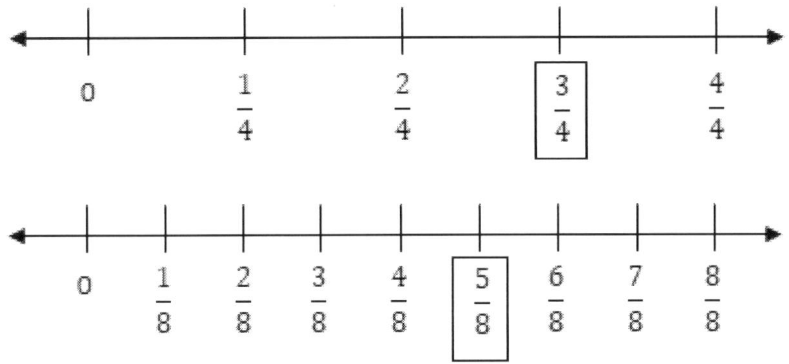

Here it can be observed that $\frac{3}{4}$ is greater than $\frac{5}{8}$, so the comparison is written as $\frac{3}{4} > \frac{5}{8}$.

This could also be shown by finding a common denominator for both fractions, so that they could be compared. First, list out factors of 4: 4, 8, 12, 16.

Then, list out factors of 8: 8, 16, 24.

Both share a common factor of 8, so they can be written in terms of 8 portions. In order for $\frac{3}{4}$ to be written in terms of 8, both the numerator and denominator must be multiplied by 2, thus forming the new fraction $\frac{6}{8}$. Now the two fractions can be compared.

Because both have the same denominator, the numerator will show the comparison.

$$\frac{6}{8} > \frac{5}{8}$$

Ordering Numbers

Whether the question asks to order the numbers from greatest to least or least to greatest, the crux of the question is the same—convert the numbers into a common format. Generally, it's easiest to write the numbers as whole numbers and decimals so they can be placed on a number line. Follow these examples to understand this strategy.

1) Order the following rational numbers from greatest to least:

$$\sqrt{36}, 0.65, 78\%, \frac{3}{4}, 7, 90\%, \frac{5}{2}$$

Of the seven numbers, the whole number (7) and decimal (0.65) are already in an accessible form, so concentrate on the other five.

First, the square root of 36 equals 6. (If the test asks for the root of a non-perfect root, determine which two whole numbers the root lies between.) Next, convert the percentages to decimals. A percentage means "per hundred," so this conversion requires moving the decimal point two places to the left, leaving 0.78 and 0.9. Lastly, evaluate the fractions:

$$\frac{3}{4} = \frac{75}{100} = 0.75$$

$$\frac{5}{2} = 2\frac{1}{2} = 2.5$$

Now, the only step left is to list the numbers in the request order:

$$7, \sqrt{36}, \frac{5}{2}, 90\%, 78\%, \frac{3}{4}, 0.65$$

2) Order the following rational numbers from least to greatest:

$$2.5, \sqrt{9}, -10.5, 0.853, 175\%, \sqrt{4}, \frac{4}{5}$$

$$\sqrt{9} = 3$$

$$175\% = 1.75$$

$$\sqrt{4} = 2$$

$$\frac{4}{5} = 0.8$$

From least to greatest, the answer is: $-10.5, \frac{4}{5}, 0.853, 175\%, \sqrt{4}, 2.5, \sqrt{9}$

It is not possible to give similar relationships between two complex numbers $a + ib$ and $c + id$. This is because the real numbers cannot be identified with the complex numbers, and there is no form of comparison between the two. However, given any polynomial equation, its solutions can be solved in the complex field. If the zeros are real, they can be written as $a + i \times 0$; if they are complex, they can be written as $a + ib$; and if they are imaginary, they can be written as ib.

Ratios and Proportions

Ratios

Ratios are used to show the relationship between two quantities. The ratio of oranges to apples in the grocery store may be 3 to 2. That means that for every 3 oranges, there are 2 apples. This comparison can be expanded to represent the actual number of oranges and apples, such as 36 oranges to 24 apples. Another example may be the number of boys to girls in a math class. If the ratio of boys to girls is given as 2 to 5, that means there are 2 boys to every 5 girls in the class. Ratios can also be compared if the units in each ratio are the same. The ratio of boys to girls in the math class can be compared to the ratio of boys to girls in a science class by stating which ratio is higher and which is lower.

Rates are used to compare two quantities with different units. *Unit rates* are the simplest form of rate. With **unit rates**, the denominator in the comparison of two units is one. For example, if someone can

type at a rate of 1,000 words in 5 minutes, then their unit rate for typing is $\frac{1,000}{5} = 200$ words in one minute or 200 words per minute. Any rate can be converted into a unit rate by dividing to make the denominator one. 1,000 words in 5 minutes has been converted into the unit rate of 200 words per minute.

Ratios and rates can be used together to convert rates into different units. For example, if someone is driving 50 kilometers per hour, that rate can be converted into miles per hour by using a ratio known as the **conversion factor.** Since the given value contains kilometers and the final answer needs to be in miles, the ratio relating miles to kilometers needs to be used. There are 0.62 miles in 1 kilometer. This, written as a ratio and in fraction form, is $\frac{0.62\ miles}{1\ km}$. To convert 50km/hour into miles per hour, the following conversion needs to be set up:

$$\frac{50\ km}{hour} * \frac{0.62\ miles}{1\ km} = 31\ miles\ per\ hour$$

The ratio between two similar geometric figures is called the **scale factor**. For example, a problem may depict two similar triangles, A and B. The scale factor from the smaller triangle A to the larger triangle B is given as 2 because the length of the corresponding side of the larger triangle, 16, is twice the corresponding side on the smaller triangle, 8. This scale factor can also be used to find the value of a missing side, x, in triangle A. Since the scale factor from the smaller triangle (A) to larger one (B) is 2, the larger corresponding side in triangle B (given as 25) can be divided by 2 to find the missing side in A (x = 12.5). The scale factor can also be represented in the equation $2A = B$ because two times the lengths of A gives the corresponding lengths of B. This is the idea behind similar triangles.

Proportions

Much like a scale factor can be written using an equation like $2A = B$, a **relationship** is represented by the equation $Y = kX$. X and Y are proportional because as values of X increase, the values of Y also increase. A relationship that is inversely proportional can be represented by the equation $Y = \frac{k}{X}$, where the value of Y decreases as the value of X increases and vice versa.

Proportional reasoning can be used to solve problems involving ratios, percentages, and averages. Ratios can be used in setting up proportions and solving them to find unknowns. For example, if a student completes an average of 10 pages of math homework in 3 nights, how long would it take the student to complete 22 pages? Both ratios can be written as fractions. The second ratio would contain the unknown.

The following proportion represents this problem, where x is the unknown number of nights:

$$\frac{10\ pages}{3\ nights} = \frac{22\ pages}{x\ nights}$$

Solving this proportion entails cross-multiplying (multiplying both sets of numbers that are diagonally across and setting them equal to each other) and results in the following equation: $10x = 22 * 3$. Simplifying and solving for x results in the exact solution: $x = 6.6\ nights$. The result would be rounded up to 7 because the homework would actually be completed on the 7th night.

The following problem uses ratios involving percentages:

If 20% of the class is girls and 30 students are in the class, how many girls are in the class?

To set up this problem, it is helpful to use the common proportion: $\frac{\%}{100} = \frac{is}{of}$. Within the proportion, % is the percentage of girls, 100 is the total percentage of the class, *is* is the number of girls, and *of* is the total number of students in the class. Most percentage problems can be written using this language. To solve this problem, the proportion should be set up as $\frac{20}{100} = \frac{x}{30}$ and then solved for x. Cross-multiplying results in the equation $20 * 30 = 100x$, which results in the solution $x = 6$. There are 6 girls in the class.

Problems involving volume, length, and other units can also be solved using ratios. For example, A problem may ask for the volume of a cone that has a radius, $r = 7m$ and a height, $h = 16m$. Referring to the formulas provided on the test, the volume of a cone is given as: $V = \pi r^2 \frac{h}{3}$, where r is the radius, and h is the height. Plugging $r = 7$ and $h = 16$ into the formula, the following is obtained:

$$V = \pi(7^2)\frac{16}{3}$$

Therefore, the volume of the cone is found to be approximately 821m^3. Sometimes, answers in different units are sought. If this problem wanted the answer in liters, 821m^3 would need to be converted.

Using the equivalence statement 1m^3 = 1,000L, the following ratio would be used to solve for liters:

$$821\text{m}^3 * \frac{1{,}000L}{1m^3}$$

Cubic meters in the numerator and denominator cancel each other out, and the answer is converted to 821,000 liters, or $8.21 * 10^5$ L.

Other conversions can also be made between different given and final units. If the temperature in a pool is 30°C, what is the temperature of the pool in degrees Fahrenheit? To convert these units, an equation is used relating Celsius to Fahrenheit. The following equation is used:

$$T_{°F} = 1.8T_{°C} + 32$$

Plugging in the given temperature and solving the equation for T yields the result:

$$T_{°F} = 1.8(30) + 32 = 86°F$$

Both units in the metric system and U.S. customary system are widely used.

Here are some more examples of how to solve for proportions:

1) $\frac{75\%}{90\%} = \frac{25\%}{x}$

To solve for x, the fractions must be cross multiplied: ($75\%x = 90\% \times 25\%$). To make things easier, let's convert the percentages to decimals: ($0.9 \times 0.25 = 0.225 = 0.75x$). To get rid of x's coefficient, each side must be divided by that same coefficient to get the answer $x = 0.3$. The question could ask for the answer as a percentage or fraction in lowest terms, which are 30% and $\frac{3}{10}$, respectively.

2) $\frac{x}{12} = \frac{30}{96}$

Cross-multiply: $96x = 30 \times 12$

Multiply: $96x = 360$

Divide: $x = 360 \div 96$

Answer: $x = 3.75$

3) $\frac{0.5}{3} = \frac{x}{6}$

Cross-multiply: $3x = 0.5 \times 6$

Multiply: $3x = 3$

Divide: $x = 3 \div 3$

Answer: $x = 1$

You may have noticed there's a faster way to arrive at the answer. If there is an obvious operation being performed on the proportion, the same operation can be used on the other side of the proportion to solve for x. For example, in the first practice problem, 75% became 25% when divided by 3, and upon doing the same to 90%, the correct answer of 30% would have been found with much less legwork. However, these questions aren't always so intuitive, so it's a good idea to work through the steps, even if the answer seems apparent from the outset.

Percentages

Think of percentages as fractions with a denominator of 100. In fact, **percentage** means "per hundred." Problems often require converting numbers from percentages, fractions, and decimals. The following explains how to work through those conversions.

Conversions

Decimals and Percentages: Since a percentage is based on "per hundred," decimals and percentages can be converted by multiplying or dividing by 100. Practically speaking, this always amounts to moving the decimal point two places to the right or left, depending on the conversion. To convert a percentage to a decimal, move the decimal point two places to the left and remove the % sign. To convert a decimal to a percentage, move the decimal point two places to the right and add a % sign. Here are some examples:

65% = 0.65
0.33 = 33%
0.215 = 21.5%
99.99% = 0.9999
500% = 5.00
7.55 = 755%

Fractions and Percentages: Remember that a percentage is a number per one hundred. So, a percentage can be converted to a fraction by making the number in the percentage the numerator and putting 100 as the denominator:

$$43\% = \frac{43}{100}$$

$$97\% = \frac{97}{100}$$

$$4.7\% = \frac{47}{1{,}000}$$

Note in the last example, that the decimal can be removed by going from 100 to 1,000, because it's accomplished by multiplying the numerator and denominator by 10.

Note that the percent symbol (%) kind of looks like a 0, a 1, and another 0. So, think of a percentage like 54% as 54 over 100. Note that it's often good to simplify a fraction into the smallest possible numbers. So, 54/100 would then become 27/50:

$$\frac{54}{100} \div \frac{2}{2} = \frac{27}{50}$$

To convert a fraction to a percent, follow the same logic. If the fraction happens to have 100 in the denominator, you're in luck. Just take the numerator and add a percent symbol:

$$\frac{28}{100} = 28\%$$

Another option is to make the denominator equal to 100. Be sure to multiply the numerator and the denominator by the same number. For example:

$$\frac{3}{20} \times \frac{5}{5} = \frac{15}{100}$$

$$\frac{15}{100} = 15\%$$

If neither of those strategies work, divide the numerator by the denominator to get a decimal:

$$\frac{9}{12} = 0.75$$

Then convert the decimal to a percentage:

$$0.75 = 75\%$$

Percent Formula

The percent formula looks like this:

$$\frac{\boldsymbol{part}}{\boldsymbol{whole}} = \frac{\%}{\mathbf{100}}$$

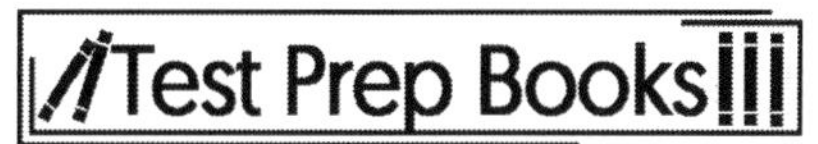

After numbers are plugged in, multiply the diagonal numbers and then divide by the remaining one. It works every time.

So, when a question asks what percent 5 is of 10. You plug the numbers in:

$$\frac{\mathbf{5}}{\mathbf{10}} = \frac{\%}{\mathbf{100}}$$

Multiply the diagonal numbers:

$$5 \times 100 = 500$$

Divide by the remaining number:

$$\frac{500}{10} = 50\%$$

The percent formula can be applied in a number of different circumstances by plugging in the numbers appropriately.

Unit Rates

Unit rate word problems will ask you to calculate the rate or quantity of something in a different value. For example, a problem might say that a car drove a certain number of miles in a certain number of minutes and then ask how many miles per hour the car was traveling. These questions involve solving proportions. Consider the following examples:

1) Alexandra made $96 during the first 3 hours of her shift as a temporary worker at a law office. She will continue to earn money at this rate until she finishes in 5 more hours. How much does Alexandra make per hour? How much will Alexandra have made at the end of the day?

This problem can be solved in two ways. The first is to set up a proportion, as the rate of pay is constant. The second is to determine her hourly rate, multiply the 5 hours by that rate, and then add the $96.

To set up a proportion, put the money already earned over the hours already worked on one side of an equation. The other side has x over 8 hours (the total hours worked in the day). It looks like this:

$$\frac{96}{3} = \frac{x}{8}$$

Now, cross-multiply to get $768 = 3x$. To get x, divide by 3, which leaves $x = 256$. Alternatively, as x is the numerator of one of the proportions, multiplying by its denominator will reduce the solution by one step. Thus, Alexandra will make $256 at the end of the day. To calculate her hourly rate, divide the total by 8, giving $32 per hour.

Alternatively, it is possible to figure out the hourly rate by dividing $96 by 3 hours to get $32 per hour. Now her total pay can be figured by multiplying $32 per hour by 8 hours, which comes out to $256.

2) Jonathan is reading a novel. So far, he has read 215 of the 335 total pages. It takes Jonathan 25 minutes to read 10 pages, and the rate is constant. How long does it take Jonathan to read one page? How much longer will it take him to finish the novel? Express the answer in time.

To calculate how long it takes Jonathan to read one page, divide the 25 minutes by 10 pages to determine the page per minute rate. Thus, it takes 2.5 minutes to read one page.

Jonathan must read 120 more pages to complete the novel. (This is calculated by subtracting the pages already read from the total.) Now, multiply his rate per page by the number of pages. Thus:

$$12 \times 2.5 = 300$$

Expressed in time, 300 minutes is equal to 5 hours.

3) At a hotel, $\frac{4}{5}$ of the 120 rooms are booked for Saturday. On Sunday, $\frac{3}{4}$ of the rooms are booked. On which day are more of the rooms booked, and by how many more?

The first step is to calculate the number of rooms booked for each day. Do this by multiplying the fraction of the rooms booked by the total number of rooms.

Saturday: $\frac{4}{5} \times 120 = \frac{4}{5} \times \frac{120}{1} = \frac{480}{5} = 96$ rooms

Sunday: $\frac{3}{4} \times 120 = \frac{3}{4} \times \frac{120}{1} = \frac{360}{4} = 90$ rooms

Thus, more rooms were booked on Saturday by 6 rooms.

4) In a veterinary hospital, the veterinarian-to-pet ratio is 1:9. The ratio is always constant. If there are 45 pets in the hospital, how many veterinarians are currently in the veterinary hospital?

Set up a proportion to solve for the number of veterinarians: $\frac{1}{9} = \frac{x}{45}$

Cross-multiplying results in $9x = 45$, which works out to 5 veterinarians.

Alternatively, as there are always 9 times as many pets as veterinarians, it is possible to divide the number of pets (45) by 9. This also arrives at the correct answer of 5 veterinarians.

5) At a general practice law firm, 30% of the lawyers work solely on tort cases. If 9 lawyers work solely on tort cases, how many lawyers work at the firm?

First, solve for the total number of lawyers working at the firm, which will be represented here with x. The problem states that 9 lawyers work solely on torts cases, and they make up 30% of the total lawyers at the firm. Thus, 30% multiplied by the total, x, will equal 9. Written as equation, this is: $30\% \times x = 9$.

It's easier to deal with the equation after converting the percentage to a decimal, leaving $0.3x = 9$. Thus:

$$x = \frac{9}{0.3} = 30$$

lawyers working at the firm.

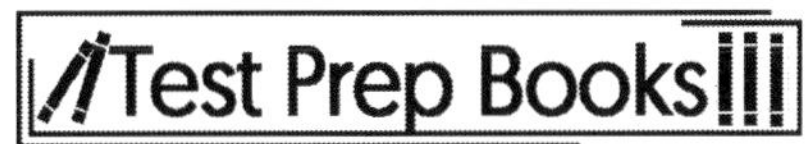

6) Xavier was hospitalized with pneumonia. He was originally given 35mg of antibiotics. Later, after his condition continued to worsen, Xavier's dosage was increased to 60mg. What was the percent increase of the antibiotics? Round the percentage to the nearest tenth.

An increase or decrease in percentage can be calculated by dividing the difference in amounts by the original amount and multiplying by 100. Written as an equation, the formula is:

$$\frac{new\ quantity\ -old\ quantity}{old\ quantity} \times 100$$

Here, the question states that the dosage was increased from 35mg to 60mg, so these are plugged into the formula to find the percentage increase.

$$\frac{60\ -\ 35}{35} \times 100\ =\ \frac{25}{35} \times 100 =\ .7142\ \times 100\ = 71.4\%$$

Multiple-Step Problems that Use Ratios, Proportions, and Percentages

Solving Real-World Problems Involving Ratios and Rates of Change

Ratios are used to show the relationship between two quantities. The ratio of oranges to apples in the grocery store may be 3 to 2. That means that for every 3 oranges, there are 2 apples. This comparison can be expanded to represent the actual number of oranges and apples, such as 36 oranges to 24 apples. Another example may be the number of boys to girls in a math class. If the ratio of boys to girls is given as 2 to 5, that means there are 2 boys to every 5 girls in the class. Ratios can also be compared if the units in each ratio are the same. The ratio of boys to girls in the math class can be compared to the ratio of boys to girls in a science class by stating which ratio is higher and which is lower.

Rates are used to compare two quantities with different units. **Unit rates** are the simplest form of rate. With unit rates, the denominator in the comparison of two units is one. For example, if someone can type at a rate of 1,000 words in 5 minutes, then their unit rate for typing is $\frac{1{,}000}{5} = 200$ words in one minute or 200 words per minute. Any rate can be converted into a unit rate by dividing to make the denominator one. 1,000 words in 5 minutes has been converted into the unit rate of 200 words per minute.

Ratios and rates can be used together to convert rates into different units. For example, if someone is driving 50 kilometers per hour, that rate can be converted into miles per hour by using a ratio known as the **conversion factor**. Since the given value contains kilometers and the final answer needs to be in miles, the ratio relating miles to kilometers needs to be used. There are 0.62 miles in 1 kilometer. This, written as a ratio and in fraction form, is $\frac{0.62\ miles}{1\ km}$. To convert 50km/hour into miles per hour, the following conversion needs to be set up:

$$\frac{50\ km}{hour} \times \frac{0.62\ miles}{1\ km} = 31\ miles\ per\ hour$$

When dealing with word problems, there is no fixed series of steps to follow, but there are some general guidelines to use. It is important that the quantity to be found is identified. Then, it can be determined how the given values can be used and manipulated to find the final answer.

Example: Jana wants to travel to visit Alice, who lives one hundred and fifty miles away. If she can drive at fifty miles per hour, how long will her trip take?

The quantity to find is the *time* of the trip. The time of a trip is given by the distance to travel divided by the speed to be traveled. The problem determines that the distance is one hundred and fifty miles, while the speed is fifty miles per hour. Thus, 150 divided by 50 is $150 \div 50 = 3$. Because *miles* and *miles per hour* are the units being divided, the miles cancel out. The result is 3 hours.

Example: Bernard wishes to paint a wall that measures twenty feet wide by eight feet high. It costs ten cents to paint one square foot. How much money will Bernard need for paint?

The final quantity to compute is the *cost* to paint the wall. This will be ten cents ($0.10) for each square foot of area needed to paint. The area to be painted is unknown, but the dimensions of the wall are given; thus, it can be calculated.

The dimensions of the wall are 20 feet wide and 8 feet high. Since the area of a rectangle is length multiplied by width, the area of the wall is:

$8 \times 20 = 160\ square\ feet$

Multiplying 0.1×160 yields $16 as the cost of the paint.

Solving Real-World Problems Involving Proportions

Much like a scale factor can be written using an equation like $2A = B$, a **relationship** is represented by the equation $Y = kX$. X and Y are proportional because as values of X increase, the values of Y also increase. A relationship that is inversely proportional can be represented by the equation $Y = \frac{k}{X}$, where the value of Y decreases as the value of X increases and vice versa.

Proportional reasoning can be used to solve problems involving ratios, percentages, and averages. Ratios can be used in setting up proportions and solving them to find unknowns. For example, if a student completes an average of 10 pages of math homework in 3 nights, how long would it take the student to complete 22 pages? Both ratios can be written as fractions. The second ratio would contain the unknown.

The following proportion represents this problem, where x is the unknown number of nights:

$$\frac{10\ pages}{3\ nights} = \frac{22\ pages}{x\ nights}$$

Solving this proportion entails cross-multiplying and results in the following equation: $10x = 22 \times 3$. Simplifying and solving for x results in the exact solution: $x = 6.6\ nights$. The result would be rounded up to 7 because the homework would actually be completed on the 7th night.

The following problem uses ratios involving percentages:

If 20% of the class is girls and 30 students are in the class, how many girls are in the class?

To set up this problem, it is helpful to use the common proportion: $\frac{\%}{100} = \frac{is}{of}$. Within the proportion, % is the percentage of girls, 100 is the total percentage of the class, *is* is the number of girls, and *of* is the

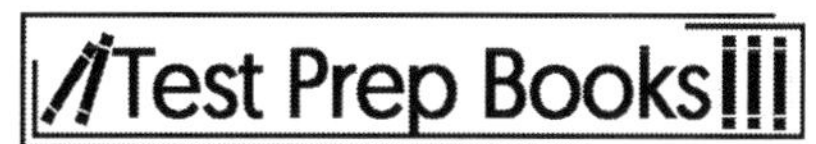

total number of students in the class. Most percentage problems can be written using this language. To solve this problem, the proportion should be set up as $\frac{20}{100} = \frac{x}{30}$, and then solved for x. Cross-multiplying results in the equation $20 \times 30 = 100x$, which results in the solution $x = 6$. There are 6 girls in the class.

Problems involving volume, length, and other units can also be solved using ratios. For example, a problem may ask for the volume of a cone to be found that has a radius:

$$r = 7m$$

and a height:

$$h = 16m$$

Referring to the formulas provided on the test, the volume of a cone is given as:

$$V = \pi r^2 \frac{h}{3}$$

where r is the radius, and h is the height. Plugging $r = 7$ and $h = 16$ into the formula, the following is obtained:

$$V = \pi(7^2)\frac{16}{3}$$

Therefore, volume of the cone is found to be $821m^3$. Sometimes, answers in different units are sought. If this problem wanted the answer in liters, $821m^3$ would need to be converted. Using the equivalence statement $1m^3$ = 1,000L, the following ratio would be used to solve for liters:

$$821\text{m}^3 \times \frac{1{,}000L}{1m^3}$$

Cubic meters in the numerator and denominator cancel each other out, and the answer is converted to 821,000 liters, or 8.21×10^5 L.

Other conversions can also be made between different given and final units. If the temperature in a pool is 30°C, what is the temperature of the pool in degrees Fahrenheit? To convert these units, an equation is used relating Celsius to Fahrenheit. The following equation is used:

$$T_{°F} = 1.8T_{°C} + 32$$

Plugging in the given temperature and solving the equation for T yields the result:

$$T_{°F} = 1.8(30) + 32 = 86°F$$

Units in both the metric system and U.S. customary system are widely used.

Here are some more examples of how to solve for proportions:

1) $\frac{75\%}{90\%} = \frac{25\%}{x}$

To solve for x, the fractions must be cross multiplied:

$$(75\%x = 90\% \times 25\%\ )$$

To make things easier, let's convert the percentages to decimals:

$$(0.9 \times 0.25 = 0.225 = 0.75x)$$

To get rid of x's coefficient, each side must be divided by that same coefficient to get the answer $x = 0.3$. The question could ask for the answer as a percentage or fraction in lowest terms, which are 30% and $\frac{3}{10}$, respectively.

2) $\frac{x}{12} = \frac{30}{96}$

Cross-multiply: $96x = 30 \times 12$

Multiply: $96x = 360$

Divide: $x = 360 \div 96$

Answer: $x = 3.75$

3) $\frac{0.5}{3} = \frac{x}{6}$

Cross-multiply: $3x = 0.5 \times 6$

Multiply: $3x = 3$

Divide: $x = 3 \div 3$

Answer: $x = 1$

You may have noticed there's a faster way to arrive at the answer. If there is an obvious operation being performed on the proportion, the same operation can be used on the other side of the proportion to solve for x. For example, in the first practice problem, 75% became 25% when divided by 3, and upon doing the same to 90%, the correct answer of 30% would have been found with much less legwork. However, these questions aren't always so intuitive, so it's a good idea to work through the steps, even if the answer seems apparent from the outset.

Solving Real-World Problems Involving Percentages

Questions dealing with percentages can be difficult when they are phrased as word problems. These word problems almost always come in three varieties. The first type will ask to find what percentage of some number will equal another number. The second asks to determine what number is some percentage of another given number. The third will ask what number another number is a given percentage of.

One of the most important parts of correctly answering percentage word problems is to identify the numerator and the denominator. This fraction can then be converted into a percentage, as described above.

The following word problem shows how to make this conversion:

A department store carries several different types of footwear. The store is currently selling 8 athletic shoes, 7 dress shoes, and 5 sandals. What percentage of the store's footwear are sandals?

First, calculate what serves as the **whole**, as this will be the denominator. How many total pieces of footwear does the store sell? The store sells 20 different types ($8\ athletic + 7\ dress + 5\ sandals$).

Second, what footwear type is the question specifically asking about? Sandals. Thus, 5 is the numerator.

Third, the resultant fraction must be expressed as a percentage. The first two steps indicate that $\frac{5}{20}$ of the footwear pieces are sandals. This fraction must now be converted into a percentage:

$$\frac{5}{20} \times \frac{5}{5} = \frac{25}{100} = 25\%$$

Simplifying Exponents

Exponents are used in mathematics to express a number or variable multiplied by itself a certain number of times. For example, x^3 means *x* is multiplied by itself three times. In this expression, x is called the **base,** and 3 is the **exponent.** Exponents can be used in more complex problems when they contain fractions and negative numbers.

Fractional exponents can be explained by looking first at the inverse of exponents, which are **roots.** Given the expression x^2, the square root can be taken, $\sqrt{x^2}$, cancelling out the 2 and leaving *x* by itself, if *x* is positive. Cancellation occurs because $\sqrt{x}$ can be written with exponents, instead of roots, as $x^{\frac{1}{2}}$. The numerator of 1 is the exponent, and the denominator of 2 is called the root (which is why it's referred to as **square root**). Taking the square root of x^2 is the same as raising it to the $\frac{1}{2}$ power. Written out in mathematical form, it takes the following progression:

$$\sqrt{x^2} = (x^2)^{\frac{1}{2}} = x$$

From properties of exponents:

$$2 \times \frac{1}{2} = 1$$

is the actual exponent of *x*. Another example can be seen with $x^{\frac{4}{7}}$. The variable *x,* raised to four-sevenths, is equal to the seventh root of *x* to the fourth power: $\sqrt[7]{x^4}$. In general,

$$x^{\frac{1}{n}} = \sqrt[n]{x}$$

and

$$x^{\frac{m}{n}} = \sqrt[n]{x^m}$$

Negative exponents also involve fractions. Whereas y^3 can also be rewritten as $\frac{y^3}{1}$, y^{-3} can be rewritten as $\frac{1}{y^3}$. A negative exponent means the exponential expression must be moved to the opposite spot in a fraction to make the exponent positive. If the negative appears in the numerator, it moves to the denominator. If the negative appears in the denominator, it is moved to the numerator. In general, $a^{-n} = \frac{1}{a^n}$, and a^{-n} and a^n are reciprocals.

Take, for example, the following expression:

$$\frac{a^{-4}b^2}{c^{-5}}$$

Since *a* is raised to the negative fourth power, it can be moved to the denominator. Since *c* is raised to the negative fifth power, it can be moved to the numerator. The *b* variable is raised to the positive second power, so it does not move.

The simplified expression is as follows:

$$\frac{b^2c^5}{a^4}$$

In mathematical expressions containing exponents and other operations, the order of operations must be followed. **PEMDAS** states that exponents are calculated after any parentheses and grouping symbols but before any multiplication, division, addition, and subtraction.

There are a few rules for working with exponents. For any numbers a, b, m, n, the following hold true:

$$a^1 = a$$

$$1^a = 1$$

$$a^0 = 1$$

$$a^m \times a^n = a^{m+n}$$

$$a^m \div a^n = a^{m-n}$$

$$(a^m)^n = a^{m \times n}$$

$$(a \times b)^m = a^m \times b^m$$

$$(a \div b)^m = a^m \div b^m$$

Any number, including a fraction, can be an exponent. The same rules apply.

Squares, Square Roots, Cubes, and Cube Roots

A **root** is a different way to write an exponent when the exponent is the reciprocal of a whole number. We use the **radical** symbol to write this in the following way:

$$\sqrt[n]{a} = a^{\frac{1}{n}}$$

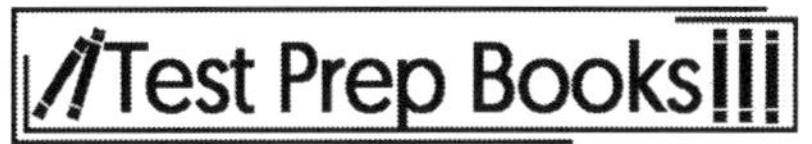

This quantity is called the *n-th* **root** of *a*. The *n* is called the **index** of the radical.

Note that if the *n*-th root of *a* is multiplied by itself *n* times, the result will just be *a*. If no number *n* is written by the radical, it is assumed that *n* is 2: $\sqrt{5} = 5^{\frac{1}{2}}$. The special case of the 2nd root is called the **square root,** and the third root is called the **cube root**.

A **perfect square** is a whole number that is the square of another whole number. For example, 16 and 64 are perfect squares because 16 is the square of 4, and 64 is the square of 8.

Undefined Expressions

Expressions can be undefined when they involve dividing by zero or having a zero denominator. In simple fractions, the numerator and denominator can be nearly any integer. However, the denominator of a fraction can never be zero because dividing by zero is a function that is undefined. Trying to take the square root of a negative number also yields an undefined result.

Objects at Scale

Scale drawings are used in designs to model the actual measurements of a real-world object. For example, the blueprint of a house might indicate that it is drawn at a scale of 3 inches to 8 feet. Given one value and asked to determine the width of the house, a proportion should be set up to solve the problem. Given the scale of 3in:8ft and a blueprint width of 1 ft (12 in.), to find the actual width of the building, the proportion $\frac{3}{8} = \frac{12}{x}$ should be used. This results in an actual width of 32 ft.

The ratio between two similar geometric figures is called the **scale factor**. For example, a problem may depict two similar triangles, A and B. The scale factor from the smaller triangle A to the larger triangle B is given as 2 because the length of the corresponding side of the larger triangle, 16, is twice the corresponding side on the smaller triangle, 8. This scale factor can also be used to find the value of a missing side, x, in triangle A. Since the scale factor from the smaller triangle (A) to larger one (B) is 2, the larger corresponding side in triangle B (given as 25) can be divided by 2 to find the missing side in A (x = 12.5). The scale factor can also be represented in the equation $2A = B$ because two times the lengths of A gives the corresponding lengths of B. This is the idea behind similar triangles.

Geometry

Side Lengths of Shapes When Given the Area or Perimeter

The **perimeter** of a polygon is the distance around the outside of the two-dimensional figure or the sum of the lengths of all the sides. Perimeter is a one-dimensional measurement and is therefore expressed in linear units such as centimeters (*cm*), feet (*ft*), and miles (*mi*). The perimeter (*P*) of a figure can be calculated by adding together each of the sides.

Properties of certain polygons allow that the perimeter may be obtained by using formulas. A regular polygon is one in which all sides have equal length and all interior angles have equal measures, such as a square and an equilateral triangle. To find the perimeter of a regular polygon, the length of one side is multiplied by the number of sides.

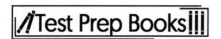

A rectangle consists of two sides called the length (*l*), which have equal measures, and two sides called the width (*w*), which have equal measures. Therefore, the perimeter (*P*) of a rectangle can be expressed as:

$$P = l + l + w + w$$

This can be simplified to produce the following formula to find the perimeter of a rectangle:

$$P = 2l + 2w \text{ or } P = 2(l + w)$$

Consider the following problem:

The total perimeter of a rectangular garden is 36m. If the length of each side is 12m, what is the width?

The formula for the perimeter of a rectangle is $P = 2L + 2W$, where P is the perimeter, L is the length, and W is the width. The first step is to substitute all of the data into the formula:

$$36 = 2(12) + 2W$$

Simplify by multiplying 2×12:

$$36 = 24 + 2W$$

Simplifying this further by subtracting 24 on each side gives:

$$36 - 24 = 24 - 24 + 2W$$

$$12 = 2W$$

Divide by 2:

$$6 = W$$

The width is 6 cm. Remember to test this answer by substituting this value into the original formula:

$$36 = 2(12) + 2(6)$$

The perimeter of a square is measured by adding together all of the sides. Since a square has four equal sides, its perimeter can be calculated by multiplying the length of one side by 4. Thus, the formula is $P = 4 \times s$, where s equals one side. For example, the following square has side lengths of 5 meters:

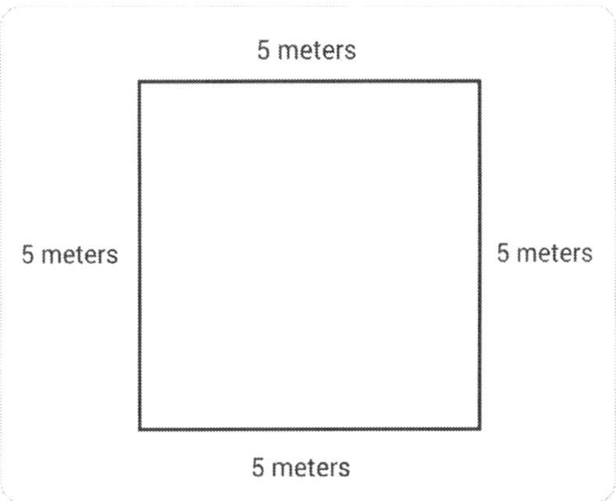

The perimeter is 20 meters because 4 times 5 is 20.

A triangle's perimeter is measured by adding together the three sides, so the formula is $P = a + b + c$, where $a, b,$ and c are the values of the three sides. The area is calculated by multiplying the length of the base times the height times ½, so the formula is:

$$A = \frac{1}{2} \times b \times h = \frac{bh}{2}$$

The base is the bottom of the triangle, and the height is the distance from the base to the peak. If a problem asks to calculate the area of a triangle, it will provide the base and height.

Missing side lengths can be determined using subtraction. For example, if you are told that a triangle has a perimeter of 34 inches and that one side is 12 inches, another side is 16 inches, and the third side is unknown, you can calculate the length of that unknown side by setting up the following subtraction problem:

$$34\ inches = 12\ inches + 16\ inches + x$$

$$34\ inches = 28\ inches + x$$

$$6\ inches = x$$

Therefore, the missing side length is 6 inches.

Area and Perimeter of Two-Dimensional Shapes

As mentioned, the **perimeter** of a polygon is the distance around the outside of the two-dimensional figure. Perimeter is a one-dimensional measurement and is therefore expressed in linear units such as centimeters (*cm*), feet (*ft*), and miles (*mi*). The perimeter (*P*) of a figure can be calculated by adding together each of the sides.

The **area** of a polygon is the number of square units needed to cover the interior region of the figure. Area is a two-dimensional measurement. Therefore, area is expressed in square units, such as square centimeters (cm^2), square feet (ft^2), or square miles (mi^2). Regarding the area of a rectangle with sides of length *x* and *y*, the area is given by xy. For a triangle with a base of length *b* and a height of length *h*, the area is $\frac{1}{2}bh$. To find the area (*A*) of a parallelogram, the length of the base (*b*) is multiplied by the length of the height:

$$(h) \rightarrow A = b \times h$$

Similar to triangles, the height of the parallelogram is measured from one base to the other at a 90° angle (or perpendicular).

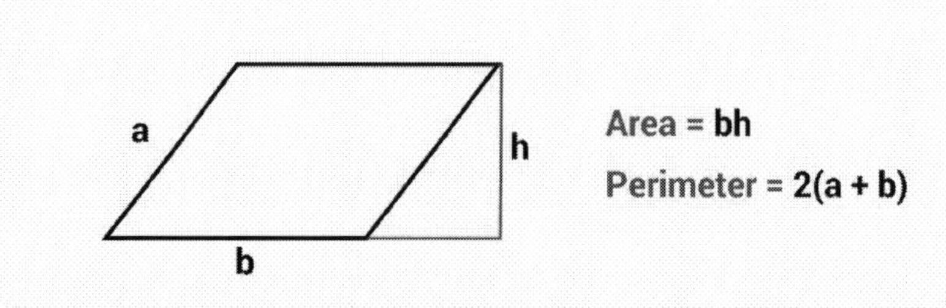

The area of a trapezoid can be calculated using the formula:

$$A = \frac{1}{2} \times h(b_1 + b_2)$$

where *h* is the height, and b_1 and b_2 are the parallel bases of the trapezoid.

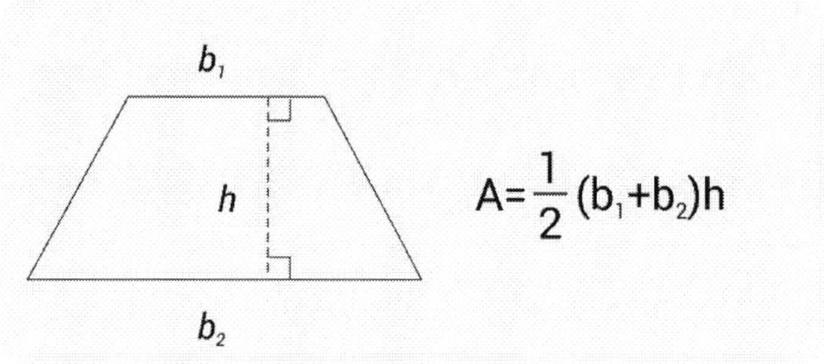

The area of a regular polygon can be determined by using its perimeter and the length of the **apothem**. The apothem is a line from the center of the regular polygon to any of its sides at a right angle. (Note that the perimeter of a regular polygon can be determined given the length of only one side.) The formula for the area (*A*) of a regular polygon is

$$A = \frac{1}{2} \times a \times P$$

where *a* is the length of the apothem, and *P* is the perimeter of the figure. Consider the following regular pentagon:

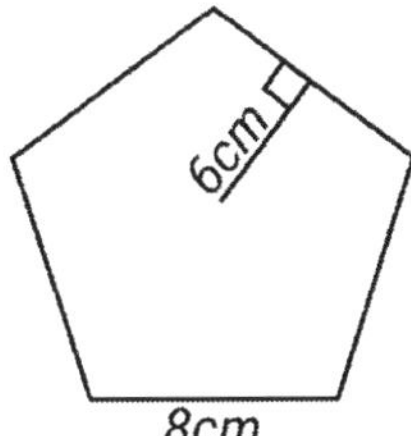

To find the area, the perimeter (*P*) is calculated first:

$$8cm \times 5 \rightarrow P = 40cm$$

Then the perimeter and the apothem are used to find the area (*A*):

$$A = \frac{1}{2} \times a \times P$$

$$A = \frac{1}{2} \times (6cm) \times (40cm)$$

$$A = 120cm^2$$

Note that the unit is:

$$cm^2 \rightarrow cm \times cm = cm^2$$

The area of irregular polygons is found by decomposing, or breaking apart, the figure into smaller shapes. When the area of the smaller shapes is determined, the area of the smaller shapes will produce the area of the original figure when added together. Consider the example below:

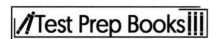

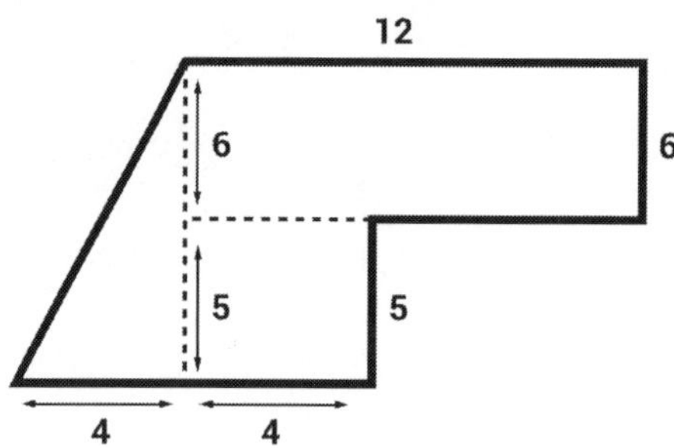

The irregular polygon is decomposed into two rectangles and a triangle. The area of the large rectangle:

$$(A = l \times w \rightarrow A = 12 \times 6)$$

is 72 square units. The area of the small rectangle is 20 square units:

$$A = 4 \times 5$$

The area of the triangle:

$$A = \frac{1}{2} \times b \times h$$

$$A = \frac{1}{2} \times 4 \times 11$$

22 square units

The sum of the areas of these figures produces the total area of the original polygon:

$$A = 72 + 20 + 22$$

A = 114 square units

The perimeter (*P*) of the figure below is calculated by:

$$P = 9m + 5m + 4m + 6m + 8m \rightarrow P = 32\ m$$

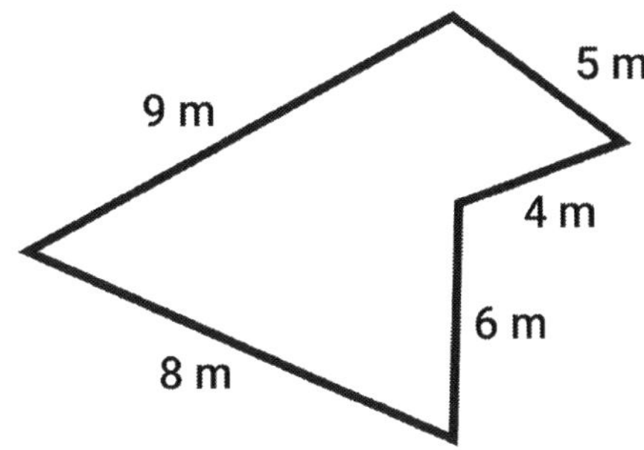

Area, Circumference, Radius, and Diameter of a Circle

A **circle** can be defined as the set of all points that are the same distance (known as the **radius**, *r*) from a single point (known as the **center** of the circle). The center has coordinates (h, k), and any point on the circle can be labelled with coordinates (x, y).

A circle's perimeter—also known as its circumference—is measured by multiplying the diameter (the straight line measured from one end to the direct opposite end of the circle) by π, so the formula is $\pi \times d$. This is sometimes expressed by the formula $C = 2 \times \pi \times r$, where r is the radius of the circle. These formulas are equivalent, as the radius equals half of the diameter.

The area of a circle is calculated through the formula $A = \pi \times r^2$. The test will indicate either to leave the answer with π attached or to calculate to the nearest decimal place, which means multiplying by 3.14 for π.

Given two points on the circumference of a circle, the path along the circle between those points is called an **arc** of the circle. For example, the arc between *B* and *C* is denoted by a thinner line:

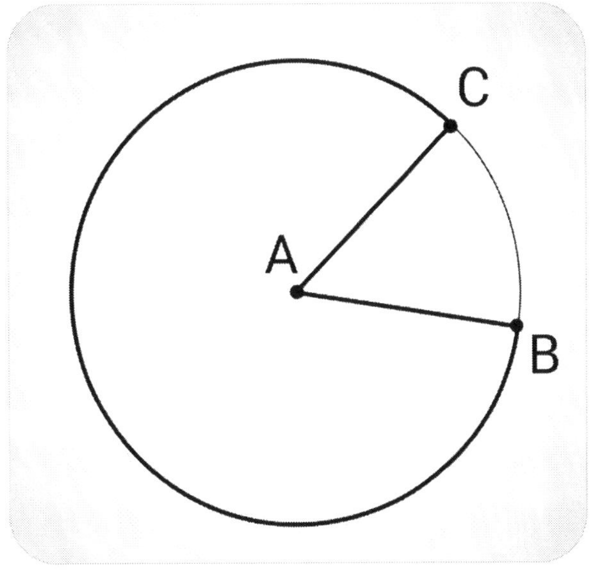

The length of the path along an arc is called the **arc length**. If the circle has radius *r*, then the arc length is given by multiplying the measure of the angle in radians by the radius of the circle.

Pythagorean Theorem

The Pythagorean theorem is an important concept in geometry. It states that for right triangles, the sum of the squares of the two shorter sides will be equal to the square of the longest side (also called the **hypotenuse**). The longest side will always be the side opposite to the 90° angle. If this side is called *c*, and the other two sides are *a* and *b*, then the Pythagorean theorem states that:

$$c^2 = a^2 + b^2$$

Since lengths are always positive, this also can be written as:

$$c = \sqrt{a^2 + b^2}$$

A diagram to show the parts of a triangle using the Pythagorean theorem is below.

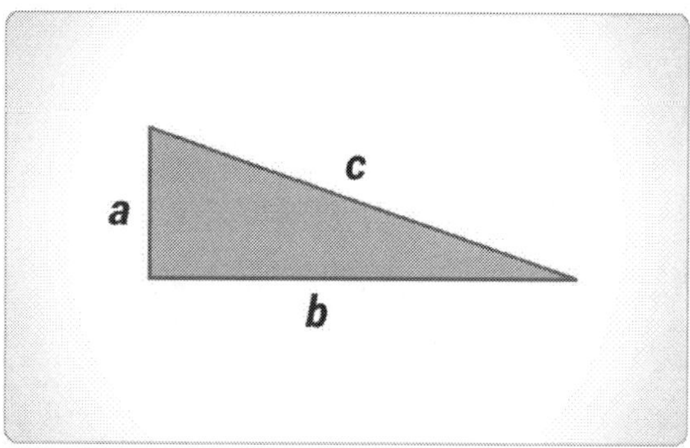

As an example of the theorem, suppose that Shirley has a rectangular field that is 5 feet wide and 12 feet long, and she wants to split it in half using a fence that goes from one corner to the opposite corner. How long will this fence need to be? To figure this out, note that this makes the field into two right triangles, whose hypotenuse will be the fence dividing it in half. Therefore, the fence length will be given by

$$\sqrt{5^2 + 12^2} = \sqrt{169} = 13 \text{ feet long}$$

Volume and Surface Area of Three-Dimensional Shapes

Geometry in three dimensions is similar to geometry in two dimensions. The main new feature is that three points now define a unique **plane** that passes through each of them. Three-dimensional objects can be made by putting together two-dimensional figures in different surfaces. Below, some of the possible three-dimensional figures will be provided, along with formulas for their volumes and surface areas.

Volume is the measurement of how much space an object occupies, like how much space is in the cube. Volume questions will ask how much of something is needed to completely fill the object. The most common surface area and volume questions deal with spheres, cubes, and rectangular prisms.

Surface area of a three-dimensional figure refers to the number of square units needed to cover the entire surface of the figure. This concept is similar to using wrapping paper to completely cover the outside of a box. For example, if a triangular pyramid has a surface area of 17 square inches (written $17in^2$), it will take 17 squares, each with sides one inch in length, to cover the entire surface of the pyramid. Surface area is also measured in square units.

A **rectangular prism** is a box whose sides are all rectangles meeting at 90° angles. Such a box has three dimensions: length, width, and height. If the length is *x*, the width is *y*, and the height is *z*, then the volume is given by $V = xyz$.

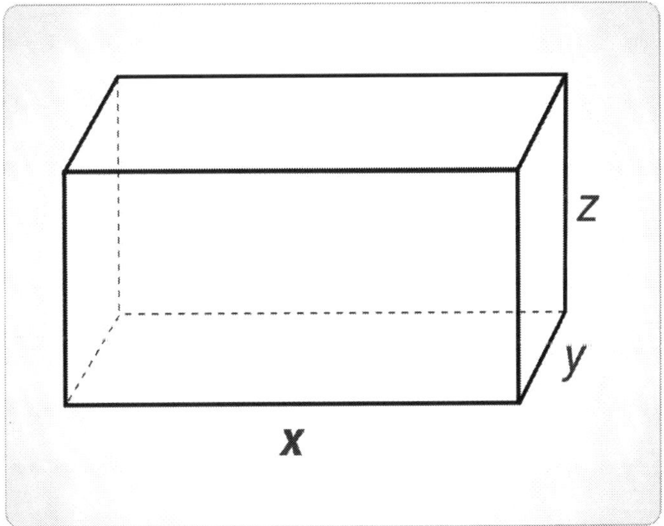

The **surface area** will be given by computing the surface area of each rectangle and adding them together. There is a total of six rectangles. Two of them have sides of length *x* and *y*, two have sides of length *y* and *z*, and two have sides of length *x* and *z*. Therefore, the total surface area will be given by:

$$SA = 2xy + 2yz + 2xz$$

A **cube** is a special type of rectangular solid in which its length, width, and height are the same. If this length is s, then the formula for the volume of a cube is $V = s \times s \times s$. The surface area of a cube is $SA = 6s^2$.

A **rectangular pyramid** is a figure with a rectangular base and four triangular sides that meet at a single vertex. If the rectangle has sides of length *x* and *y*, then the volume will be given by $V = \frac{1}{3}xyh$.

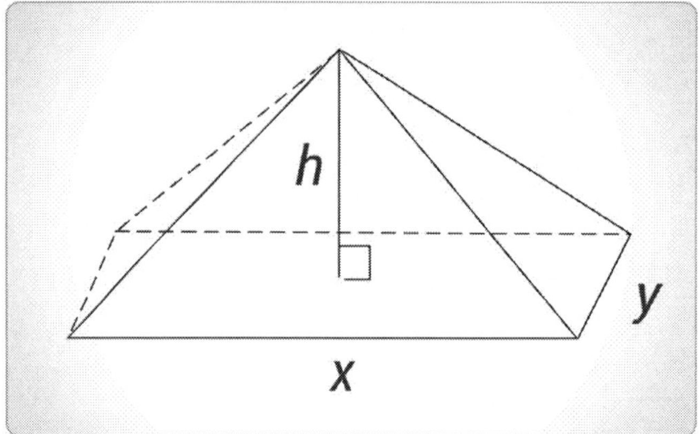

Many three-dimensional figures (solid figures) can be represented by nets consisting of rectangles and triangles. The surface area of such solids can be determined by adding the areas of each of its faces and bases. Finding the surface area using this method requires calculating the areas of rectangles and triangles.

Consider the following triangular prism, which is represented by a net consisting of two triangles and three rectangles.

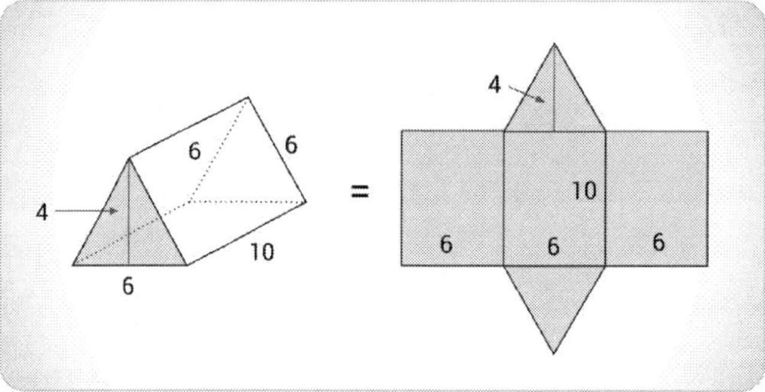

The surface area of the prism can be determined by adding the areas of each of its faces and bases. The surface area (*SA*) = area of triangle + area of triangle + area of rectangle + area of rectangle + area of rectangle.

$$SA = \left(\frac{1}{2} \times b \times h\right) + \left(\frac{1}{2} \times b \times h\right) + (l \times w) + (l \times w) + (l \times w)$$

$$SA = \left(\frac{1}{2} \times 6 \times 4\right) + \left(\frac{1}{2} \times 6 \times 4\right) + (6 \times 10) + (6 \times 10) + (6 \times 10)$$

$$SA = (12) + (12) + (60) + (60) + (60)$$

$$SA = 204 \; square \; units$$

A **sphere** is a set of points all of which are equidistant from some central point. It is like a circle, but in three dimensions. The volume of a sphere of radius *r* is given by:

$$V = \frac{4}{3}\pi r^3$$

The surface area is given by $A = 4\pi r^2$.

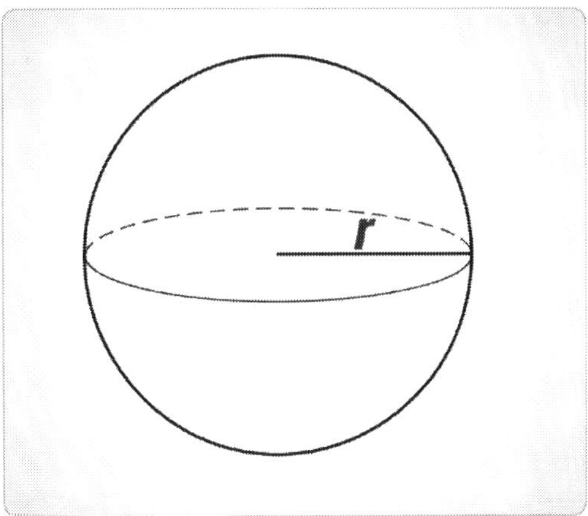

The volume of a **cylinder** is then found by adding a third dimension onto the circle. Volume of a cylinder is calculated by multiplying the area of the base (which is a circle) by the height of the cylinder. Doing so results in the equation $V = \pi r^2 h$. The volume of a **cone** is $\frac{1}{3}$ of the volume of a cylinder. Therefore, the formula for the volume of a **cone** is:

$$\frac{1}{3}\pi r^2 h$$

Solving Three-Dimensional Problems

Three-dimensional objects can be simplified into related two-dimensional shapes to solve problems. This simplification can make problem-solving a much easier experience. An isometric representation of a three-dimensional object can be completed so that important properties (e.g., shape, relationships of faces and surfaces) are noted. Edges and vertices can be translated into two-dimensional objects as well. For example, below is a three-dimensional object that's been partitioned into two-dimensional representations of its faces:

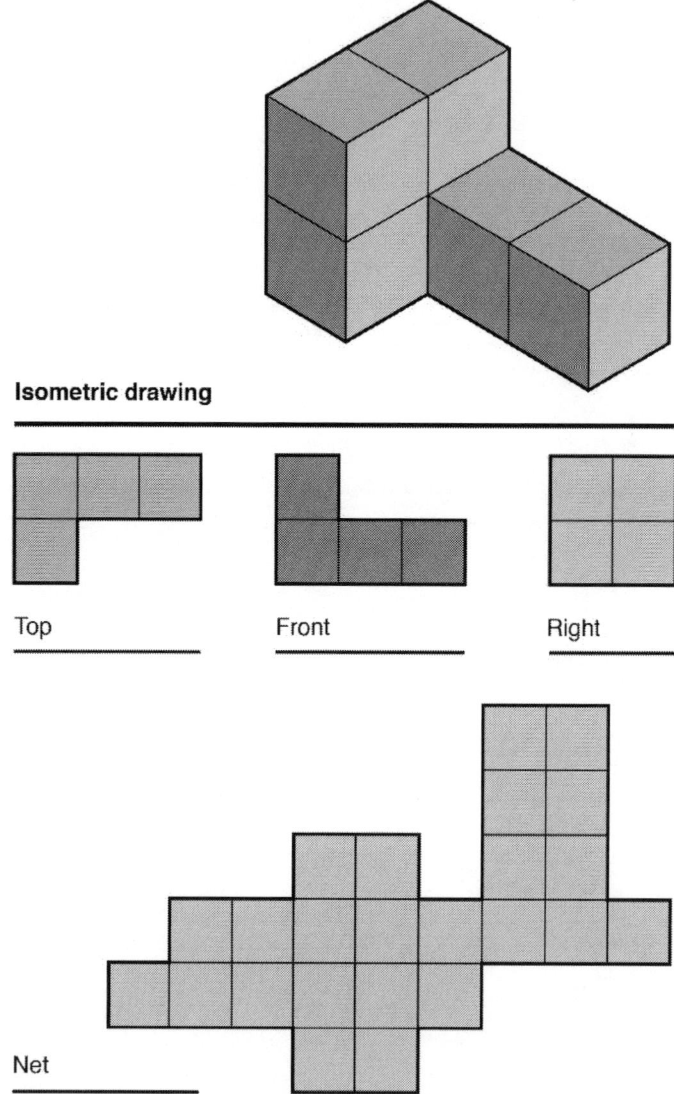

The net represents the sum of the three different faces. Depending on the problem, using a smaller portion of the given shape may be helpful, by simplifying the steps necessary to solve.

Many objects in the real world consist of three-dimensional shapes such as prisms, cylinders, and spheres. Surface area problems involve quantifying the outside area of such a three-dimensional object,

and volume problems involve quantifying how much space the object takes up. Surface area of a prism is the sum of the areas, which is simplified into $SA = 2A + Bh$, where *A* is the area of the base, *B* is the perimeter of the base, and *h* is the height of the prism. The volume of the same prism is $V = Ah$. The surface area of a cylinder is equal to the sum of the areas of each end and the side, which is:

$$SA = 2\pi rh + 2\pi r^2$$

An example when one of these formulas should be used would be when calculating how much paint is needed for the outside of a house. In this scenario, surface area must be used. The sum of all individual areas of each side of the house must be found. Also, when calculating how much water a cylindrical tank can hold, a volume formula is used. Therefore, the amount of water that a cylindrical tank that is 8 feet tall with a radius of 3 feet is:

$$\pi \times 3^2 \times 8 = 226.1 \text{ cubic feet}$$

The formula used to calculate the volume of a cone is $\frac{1}{3}\pi r^2 h$. In a real-life example where the radius of a cone is 2 meters and the height of a cone is 5 meters, the volume of the cone is calculated by utilizing the formula:

$$\frac{1}{3}\pi 2^2 \times 5$$

After substituting 3.14 for π, the volume is $20.9\ m^3$.

Graphical Data Including Graphs, Tables, and More

A set of data can be visually displayed in various forms allowing for quick identification of characteristics of the set. **Histograms**, such as the one shown below, display the number of data points (vertical axis) that fall into given intervals (horizontal axis) across the range of the set. The histogram below displays the heights of black cherry trees in a certain city park. Each rectangle represents the number of trees with heights between a given five-point span. For example, the furthest bar to the right indicates that two trees are between 85 and 90 feet. Histograms can describe the center, spread, shape, and any unusual characteristics of a data set.

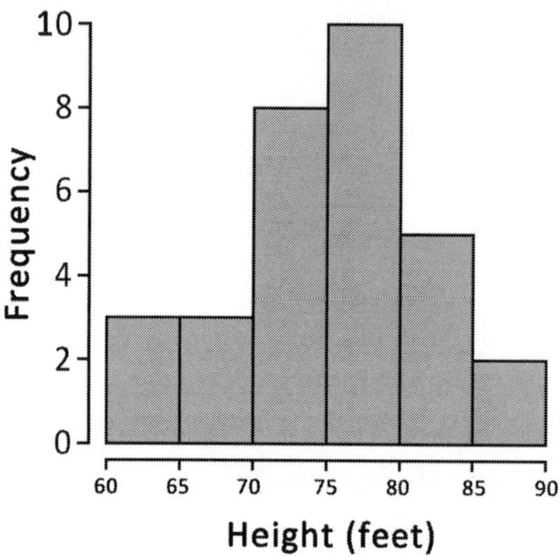

A **box plot**, also called a **box-and-whisker plot**, divides the data points into four groups and displays the five-number summary for the set as well as any outliers. The five-number summary consists of:

- The lower extreme: the lowest value that is not an outlier
- The higher extreme: the highest value that is not an outlier
- The median of the set: also referred to as the second quartile or Q_2
- The first quartile or Q_1: the median of values below Q_2
- The third quartile or Q_3: the median of values above Q_2

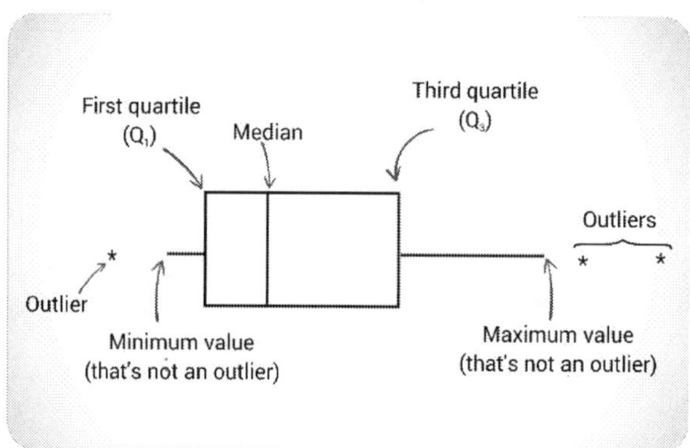

Suppose the box plot displays IQ scores for 12th grade students at a given school. The five-number summary of the data consists of: lower extreme (67); upper extreme (127); Q_2or median (100); Q_1 (91); Q_3 (108); and outliers (135 and 140). Although all data points are not known from the plot, the points are divided into four quartiles each, including 25% of the data points. Therefore, 25% of students scored between 67 and 91, 25% scored between 91 and 100, 25% scored between 100 and 108, and 25% scored between 108 and 127. These percentages include the normal values for the set and exclude the outliers. This information is useful when comparing a given score with the rest of the scores in the set.

A **scatter plot** is a mathematical diagram that visually displays the relationship or connection between two variables. The independent variable is placed on the *x*-axis, or horizontal axis, and the dependent variable is placed on the *y*-axis, or vertical axis. When visually examining the points on the graph, if the points model a linear relationship, or if a line of best-fit can be drawn through the points with the points relatively close on either side, then a correlation exists. If the line of best-fit has a positive slope (rises from left to right), then the variables have a positive correlation. If the line of best-fit has a negative slope (falls from left to right), then the variables have a negative correlation. If a line of best-fit cannot be drawn, then no correlation exists. A positive or negative correlation can be categorized as strong or weak, depending on how closely the points are graphed around the line of best-fit.

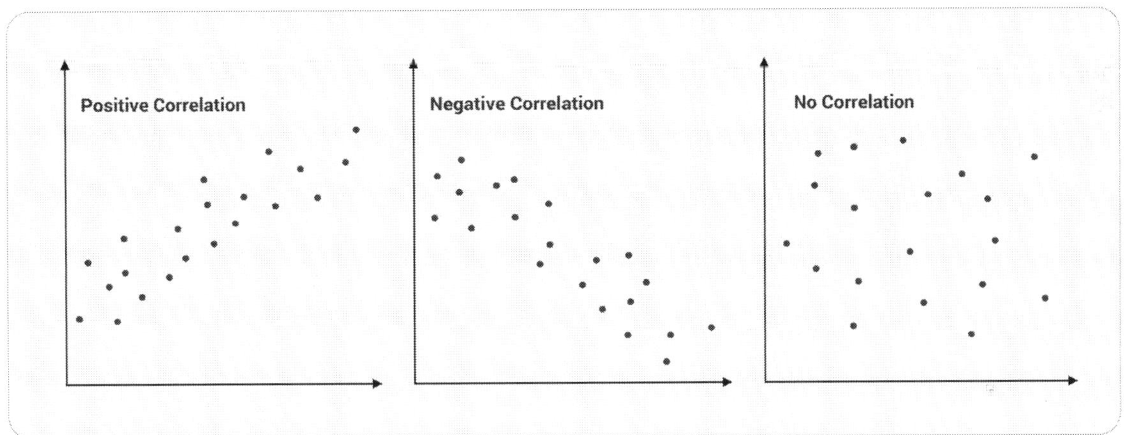

Like a scatter plot, a **line graph** compares variables that change continuously, typically over time. Paired data values (ordered pair) are plotted on a coordinate grid with the *x*- and *y*-axis representing the variables. A line is drawn from each point to the next, going from left to right. The line graph below displays cell phone use for given years (two variables) for men, women, and both sexes (three data sets).

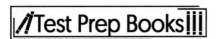

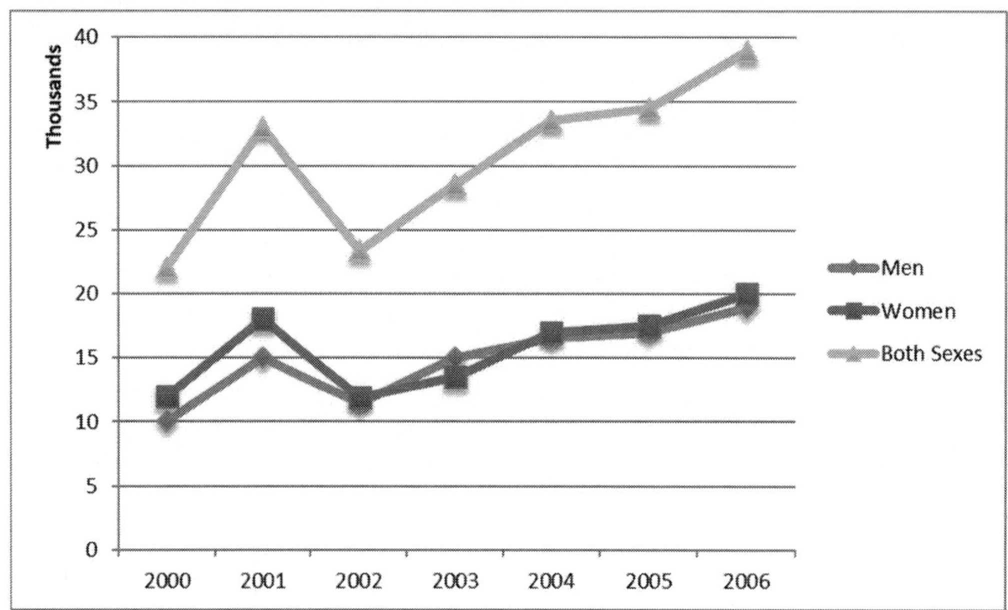

A **line plot**, also called **dot plot**, displays the frequency of data (numerical values) on a number line. To construct a line plot, a number line is used that includes all unique data values. It is marked with x's or dots above the value the number of times that the value occurs in the data set.

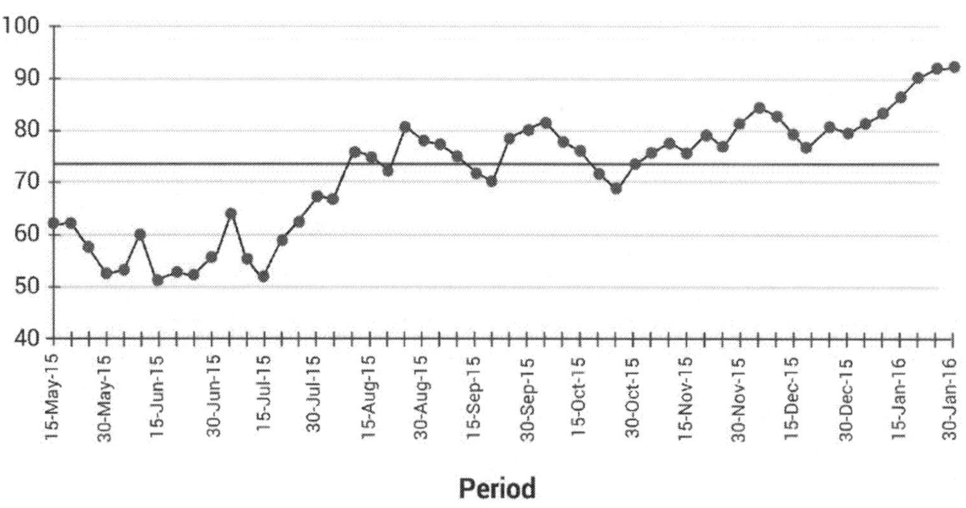

A **bar graph** is a diagram in which the quantity of items within a specific classification is represented by the height of a rectangle. Each type of classification is represented by a rectangle of equal width. Here is an example of a bar graph:

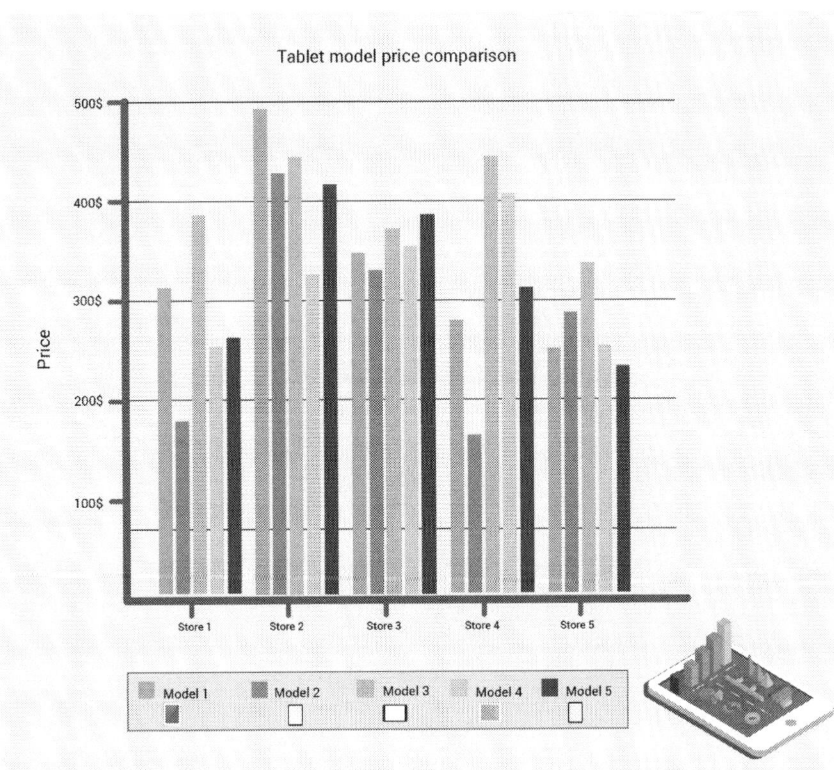

A **circle graph**, also called a **pie chart**, shows categorical data with each category representing a percentage of the whole data set. To make a circle graph, the percent of the data set for each category must be determined. To do so, the frequency of the category is divided by the total number of data points and converted to a percent. For example, if 80 people were asked what their favorite sport is and 20 responded basketball, basketball makes up 25% of the data ($\frac{20}{80} = 0.25 = 25\%$). Each category in a data set is represented by a slice of the circle proportionate to its percentage of the whole.

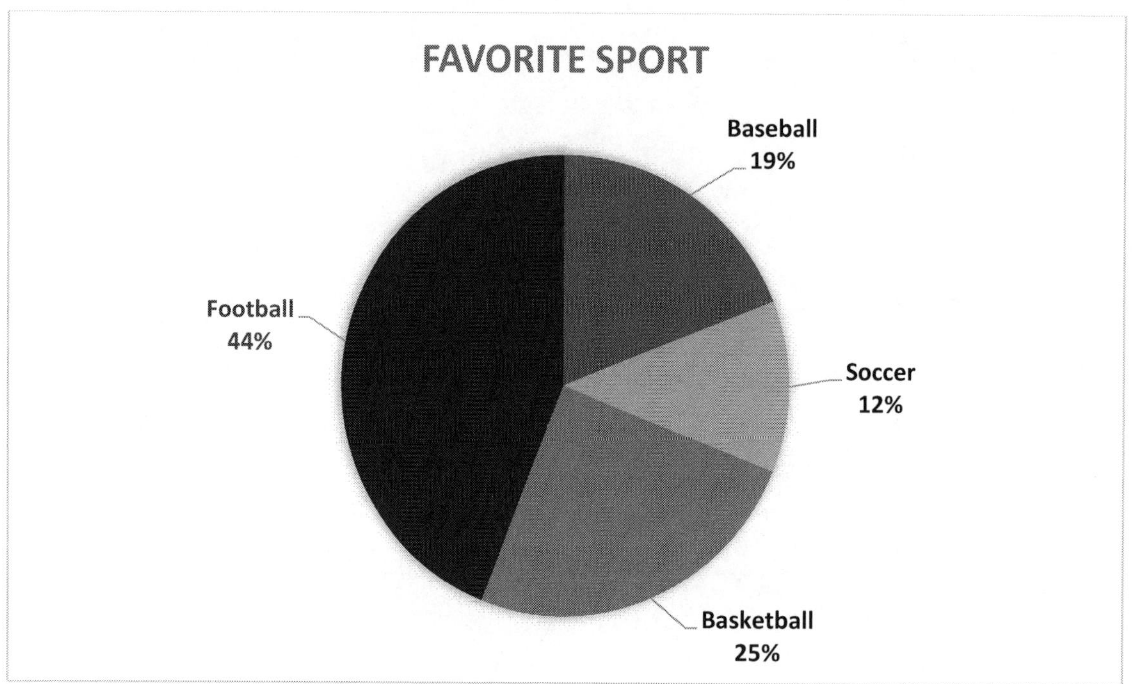

A **stem-and-leaf plot** is a method of displaying sets of data by organizing numbers by their stems (usually the tens digit) and different leaf values (usually the ones digit).

For example, to organize a number of movie critic's ratings, as listed below, a stem-and-leaf plot could be utilized to display the information in a more condensed manner.

Movie critic scores: 47, 52, 56, 59, 61, 64, 66, 68, 68, 70, 73, 75, 79, 81, 83, 85, 86, 88, 88, 89, 90, 90, 91, 93, 94, 96, 96, 99.

	Movie Ratings							
4	7							
5	2	6	9					
6	1	4	6	8	8			
7	0	3	5	9				
8	1	3	5	6	8	8	9	
9	0	0	1	3	4	6	6	9
Key	6 | 1 represents 61							

Looking at this stem and leaf plot, it is easy to ascertain key features of the data set. For example, what is the range of the data in the stem-and-leaf plot?

Using this method, it is easier to visualize the distribution of the scores and answer the question pertaining to the range of scores, which is $99 - 47 = 52$.

A **tally chart** is a diagram in which tally marks are utilized to represent data. Tally marks are a means of showing a quantity of objects within a specific classification. Here is an example of a tally chart:

Number of days with rain	Number of weeks
0	II
1	~~IIII~~
2	~~IIII~~
3	~~IIII~~
4	~~IIII~~ ~~IIII~~ ~~IIII~~ IIII
5	~~IIII~~ I
6	~~IIII~~ I
7	IIII

Data is often recorded using fractions, such as half a mile, and understanding fractions is critical because of their popular use in real-world applications. Also, it is extremely important to label values with their units when using data. For example, regarding length, the number 2 is meaningless unless it is attached to a unit. Writing 2 cm shows that the number refers to the length of an object.

Mean, Median, Mode, and Range

Suppose that *X* is a set of data points $(x_1, x_2, x_3, \dots x_n)$ and some description of the general properties of this data need to be found.

The first property that can be defined for this set of data is the **mean**. To find the mean, add up all the data points, then divide by the total number of data points. This can be expressed using **summation notation** as:

$$\bar{X} = \frac{x_1 + x_2 + x_3 + \dots + x_n}{n} = \frac{1}{n}\sum_{i=1}^{n} x_i$$

For example, suppose that in a class of 10 students, the scores on a test were 50, 60, 65, 65, 75, 80, 85, 85, 90, 100. Therefore, the average test score will be:

$$\frac{1}{10}(50 + 60 + 65 + 65 + 75 + 80 + 85 + 85 + 90 + 100) = 75.5$$

The mean is a useful number if the distribution of data is normal (more on this later), which roughly means that the frequency of different outcomes has a single peak and is roughly equally distributed on both sides of that peak. However, it is less useful in some cases where the data might be split or where there are some outliers. **Outliers** are data points that are far from the rest of the data. For example, suppose there are 10 executives and 90 employees at a company. The executives make $1,000 per hour, and the employees make $10 per hour.

Therefore, the average pay rate will be:

$$\frac{\$1{,}000 \times 10 + \$10 \times 90}{100} = \$109 \text{ per hour}$$

In this case, this average is not very descriptive since it's not close to the actual pay of the executives or the employees.

Another useful measurement is the **median**. In a data set *X* consisting of data points $x_1, x_2, x_3, \ldots x_n$, the median is the point in the middle. The middle refers to the point where half the data comes before it and half comes after, when the data is recorded in numerical order. If *n* is odd, then the median is:

$$x_{\frac{n+1}{2}}$$

If *n* is even, it is defined as $\frac{1}{2}\left(x_{\frac{n}{2}} + x_{\frac{n}{2}+1}\right)$, the mean of the two data points closest to the middle of the data points. In the previous example of test scores, the two middle points are 75 and 80. Since there is no single point, the average of these two scores needs to be found. The average is:

$$\frac{75 + 80}{2} = 77.5$$

The median is generally a good value to use if there are a few outliers in the data. It prevents those outliers from affecting the "middle" value as much as when using the mean.

One additional measure to define for *X* is the **mode**. This is the data point that appears more frequently. If two or more data points all tie for the most frequent appearance, then each of them is considered a mode. In the case of the test scores, where the numbers were 50, 60, 65, 65, 75, 80, 85, 85, 90, 100, there are two modes: 65 and 85.

Since an outlier is a data point that is far from most of the other data points in a data set, this means an outlier also is any point that is far from the median of the data set. The outliers can have a substantial effect on the mean of a data set but usually do not change the median or mode, or do not change them by a large quantity. For example, consider the data set (3, 5, 6, 6, 6, 8). This has a median of 6 and a mode of 6, with a mean of $\frac{34}{6} \approx 5.67$. Now, suppose a new data point of 1,000 is added so that the data set is now (3, 5, 6, 6, 6, 8, 1,000). The median and mode, which are both still 6, remain unchanged. However, the average is now $\frac{1034}{7}$, which is approximately 147.7. In this case, the median and mode will be better descriptions for most of the data points.

Outliers in a given data set are sometimes the result of an error by the experimenter, but oftentimes, they are perfectly valid data points that must be taken into consideration.

The **first quartile** of a set of data *X* refers to the largest value from the first ¼ of the data points. In practice, there are sometimes slightly different definitions that can be used, such as the median of the first half of the data points (excluding the median itself if there are an odd number of data points). The term also has a slightly different use: when it is said that a data point lies in the first quartile, it means it is less than or equal to the median of the first half of the data points. Conversely, if it lies *at* the first quartile, then it is equal to the first quartile.

When it is said that a data point lies in the **second quartile**, it means it is between the first quartile and the median.

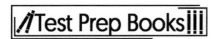

The **third quartile** refers to data that lies between ½ and ¾ of the way through the data set. Again, there are various methods for defining this precisely, but the simplest way is to include all of the data that lie between the median and the median of the top half of the data.

Data that lies in the **fourth quartile** refers to all of the data above the third quartile.

Percentiles may be defined in a similar manner to quartiles. Generally, this is defined in the following manner:

If a data point lies *in* the n-th percentile, this means it lies in the range of the first *n*% of the data.

If a data point lies *at* the *n*-th percentile, then it means that *n*% of the data lies below this data point.

Given a data set *X* consisting of data points $(x_1, x_2, x_3, \ldots x_n)$, the **variance of X** is defined to be:

$$\frac{\sum_{i=1}^{n}(x_i - \bar{X})^2}{n}$$

This means that the variance of *X* is the average of the squares of the differences between each data point and the mean of *X*. In the formula, $\bar{X}$ is the mean of the values in the data set, and x_i represents each individual value in the data set. The sigma notation indicates that the sum should be found with n being the number of values to add together. $i = 1$ means that the values should begin with the first value.

Given a data set *X* consisting of data points $(x_1, x_2, x_3, \ldots x_n)$, the **standard deviation of X** is defined to be

$$s_x = \sqrt{\frac{\sum_{i=1}^{n}(x_i - \bar{X})^2}{n}}$$

In other words, the standard deviation is the square root of the variance.

Both the variance and the standard deviation are measures of how much the data tend to be spread out. When the standard deviation is low, the data points are mostly clustered around the mean. When the standard deviation is high, this generally indicates that the data are quite spread out, or else that there are a few substantial outliers.

As a simple example, compute the standard deviation for the data set (1, 3, 3, 5). First, compute the mean, which will be:

$$\frac{1 + 3 + 3 + 5}{4} = \frac{12}{4} = 3$$

Now, find the variance of *X* with the formula:

$$\sum_{i=1}^{4}(x_i - \bar{X})^2 = (1-3)^2 + (3-3)^2 + (3-3)^2 + (5-3)^2$$

$$-2^2 + 0^2 + 0^2 + 2^2 = 8$$

Therefore, the variance is $\frac{8}{4} = 2$. Taking the square root, the standard deviation will be $\sqrt{2}$.

Note that the standard deviation only depends upon the mean, not upon the median or mode(s). Generally, if there are multiple modes that are far apart from one another, the standard deviation will be high. A high standard deviation does not always mean there are multiple modes, however.

Describing a Set of Data

A set of data can be described in terms of its center, spread, shape and any unusual features. The center of a data set can be measured by its mean, median, or mode. The spread of a data set refers to how far the data points are from the center (mean or median). The spread can be measured by the range or the quartiles and interquartile range. A data set with all its data points clustered around the center will have a small spread. A data set covering a wide range of values will have a large spread.

When a data set is displayed as a **histogram** or frequency distribution plot, the shape indicates if a sample is normally distributed, symmetrical, or has measures of skewness or kurtosis. When graphed, a data set with a **normal distribution** will resemble a bell curve.

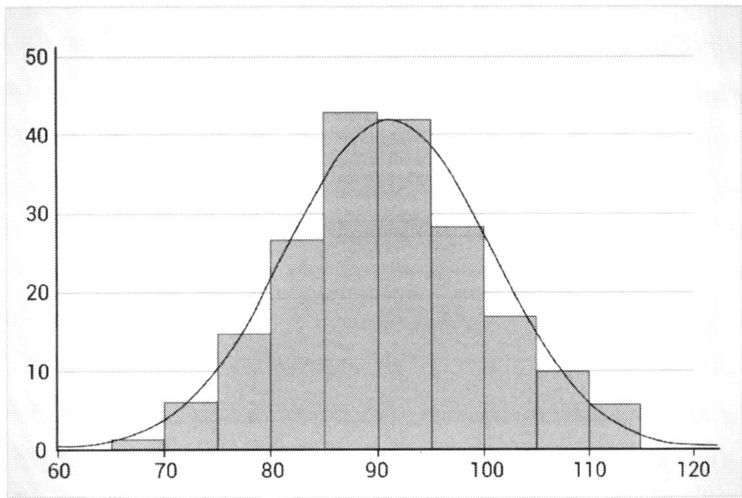

If the data set is symmetrical, each half of the graph when divided at the center is a mirror image of the other. If the graph has fewer data points to the right, the data is **skewed right**. If it has fewer data points to the left, the data is **skewed left**.

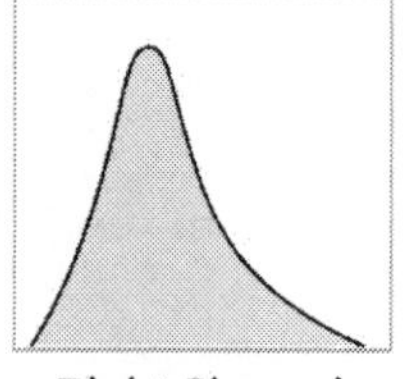
Right-Skewed

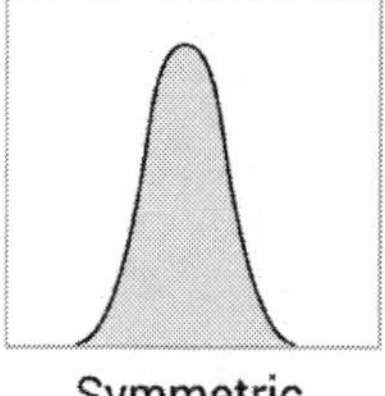
Symmetric

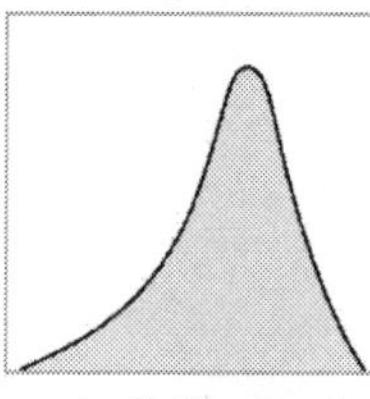
Left-Skewed

Kurtosis is a measure of whether the data is heavy-tailed with a high number of outliers, or light-tailed with a low number of outliers.

A description of a data set should include any unusual features such as gaps or outliers. A **gap** is a span within the range of the data set containing no data points. An **outlier** is a data point with a value either extremely large or extremely small when compared to the other values in the set.

Counting Techniques

The **addition rule** for probabilities states that the probability of *A* or *B* happening is:

$$P(A \cup B) = P(A) + P(B) - P(A \cap B)$$

Note that the subtraction of $P(A \cap B)$ must be performed, or else it would result in double counting any outcomes that lie in both *A* and in *B*. For example, suppose that a 20-sided die is being rolled. Fred bets that the outcome will be greater than 10, while Helen bets that it will be greater than 4 but less than 15. What is the probability that at least one of them is correct?

We apply the rule:

$$P(A \cup B) = P(A) + P(B) - P(A \cap B)$$

where *A* is that outcome *x* is in the range $x > 10$, and *B* is that outcome *x* is in the range $4 < x < 15$.

$$P(A) = 10 \times \frac{1}{20} = \frac{1}{2}$$

$$P(B) = 10 \times \frac{1}{20} = \frac{1}{2}$$

$P(A \cap B)$ can be computed by noting that $A \cap B$ means the outcome *x* is in the range $10 < x < 15$, so

$$P(A \cap B) = 4 \times \frac{1}{20} = \frac{1}{5}$$

Therefore:

$$P(A \cup B) = P(A) + P(B) - P(A \cap B)$$

$$\frac{1}{2}+\frac{1}{2}-\frac{1}{5}=\frac{4}{5}$$

Note that in this particular example, we could also have directly reasoned about the set of possible outcomes $A \cup B$, by noting that this would mean that *x* must be in the range $5 \leq x$. However, this is not always the case, depending on the given information.

The **multiplication rule** for probabilities states the probability of *A* and *B* both happening is:

$$P(A \cap B) = P(A)P(B|A)$$

As an example, suppose that when Jamie wears black pants, there is a ½ probability that she wears a black shirt as well, and that she wears black pants ¾ of the time. What is the probability that she is wearing both a black shirt and black pants?

To figure this, use the above formula, where *A* will be "Jamie is wearing black pants," while *B* will be "Jamie is wearing a black shirt." It is known that $P(A)$ is ¾. It is also known that $P(B|A) = \frac{1}{2}$. Multiplying the two, the probability that she is wearing both black pants and a black shirt is:

$$P(A)P(B|A) = \frac{3}{4} \times \frac{1}{2} = \frac{3}{8}$$

Probability of an Event

Given a set of possible outcomes *X*, a **probability distribution** of *X* is a function that assigns a probability to each possible outcome. If the outcomes are $(x_1, x_2, x_3, \ldots x_n)$, and the probability distribution is *p*, then the following rules are applied.

- $0 \leq p(x_i) \leq 1$, for any i.
- $\sum_{i=1}^{n} p(x_i) = 1$.

In other words, the probability of a given outcome must be between zero and 1, while the total probability must be 1.

If $p(x_i)$ is constant, then this is called a **uniform probability distribution**, and $p(x_i) = \frac{1}{n}$. For example, on a six-sided die, the probability of each of the six outcomes will be $\frac{1}{6}$.

If seeking the probability of an outcome occurring in some specific range *A* of possible outcomes, written $P(A)$, add up the probabilities for each outcome in that range. For example, consider a six-sided die, and figure the probability of getting a 3 or lower when it is rolled. The possible rolls are 1, 2, 3, 4, 5, and 6. So, to get a 3 or lower, a roll of 1, 2, or 3 must be completed. The probabilities of each of these is $\frac{1}{6}$, so add these to get:

$$p(1) + p(2) + p(3) = \frac{1}{6}+\frac{1}{6}+\frac{1}{6}=\frac{1}{2}$$

An outcome occasionally lies within some range of possibilities *B*, and the probability that the outcomes also lie within some set of possibilities *A* needs to be figured. This is called a **conditional probability**. It is written as $P(A|B)$, which is read "the probability of *A* given *B*." The general formula for computing conditional probabilities is:

$$P(A|B) = \frac{P(A \cap B)}{P(B)}$$

However, when dealing with uniform probability distributions, simplify this a bit. Write $|A|$ to indicate the number of outcomes in *A*. Then, for uniform probability distributions, write:

$$P(A|B) = \frac{|A \cap B|}{|B|}$$

(recall that $A \cap B$ means "*A* intersect *B*," and consists of all of the outcomes that lie in both *A* and *B*)

This means that all possible outcomes do not need to be known. To see why this formula works, suppose that the set of outcomes *X* is $(x_1, x_2, x_3, \ldots x_n)$, so that $|X| = n$. Then, for a uniform probability distribution:

$$P(A) = \frac{|A|}{n}$$

However, this means:

$$(A|B) = \frac{P(A \cap B)}{P(B)} = \frac{\frac{|A \cap B|}{n}}{\frac{|B|}{n}} = \frac{|A \cap B|}{|B|}$$

since the *n*'s cancel out.

For example, suppose a die is rolled, and it is known that it will land between 1 and 4. However, how many sides the die has is unknown. Figure the probability that the die is rolled higher than 2. To figure this, $P(3)$ or $P(4)$ does not need to be determined, or any of the other probabilities, since it is known that a fair die has a uniform probability distribution. Therefore, apply the formula $\frac{|A \cap B|}{|B|}$. So, in this case *B* is (1, 2, 3, 4) and $A \cap B$ is (3, 4). Therefore:

$$\frac{|A \cap B|}{|B|} = \frac{2}{4} = \frac{1}{2}$$

Conditional probability is an important concept because, in many situations, the likelihood of one outcome can differ radically depending on how something else comes out. The probability of passing a test given that one has studied all of the material is generally much higher than the probability of passing a test given that one has not studied at all. The probability of a person having heart trouble is much lower if that person exercises regularly. The probability that a college student will graduate is higher when their SAT scores are higher, and so on. For this reason, there are many people who are interested in conditional probabilities.

Note that in some practical situations, changing the order of the conditional probabilities can make the outcome very different. For example, the probability that a person with heart trouble has exercised regularly is quite different than the probability that a person who exercises regularly will have heart

trouble. The probability of a person receiving a military-only award, given that he or she is or was a soldier, is generally not very high, but the probability that a person being or having been a soldier, given that he or she received a military-only award, is 1.

However, in some cases, the outcomes do not influence one another this way. If the probability of *A* is the same regardless of whether *B* is given; that is, if $P(A|B) = P(A)$, then *A* and *B* are considered **independent**. In this case:

$$P(A|B) = \frac{P(A \cap B)}{P(B)} = P(A)$$

So:

$$P(A \cap B) = P(A)P(B)$$

In fact, if $P(A \cap B) = P(A)P(B)$, it can be determined that $P(A|B) = P(A)$ and $P(A|B) = P(B)$ by working backward. Therefore, *B* is also independent of *A.*

An example of something being independent can be seen in rolling dice. In this case, consider a red die and a green die. It is expected that when the dice are rolled, the outcome of the green die should not depend in any way on the outcome of the red die. Or, to take another example, if the same die is rolled repeatedly, then the next number rolled should not depend on which numbers have been rolled previously. Similarly, if a coin is flipped, then the next flip's outcome does not depend on the outcomes of previous flips.

This can sometimes be counter-intuitive, since when rolling a die or flipping a coin, there can be a streak of surprising results. If, however, it is known that the die or coin is fair, then these results are just the result of the fact that over long periods of time, it is very likely that some unlikely streaks of outcomes will occur. Therefore, avoid making the mistake of thinking that when considering a series of independent outcomes, a particular outcome is "due to happen" simply because a surprising series of outcomes has already been seen.

There is a second type of common mistake that people tend to make when reasoning about statistical outcomes: the idea that when something of low probability happens, this is surprising. It would be surprising that something with low probability happened after just one attempt. However, with so much happening all at once, it is easy to see at least something happen in a way that seems to have a very low probability. In fact, a lottery is a good example. The odds of winning a lottery are very small, but the odds that somebody wins the lottery each week are actually fairly high. Therefore, no one should be surprised when some low probability things happen.

A **simple event** consists of only one outcome. The most popular simple event is flipping a coin, which results in either heads or tails. A **compound event** results in more than one outcome and consists of more than one simple event. An example of a compound event is flipping a coin while tossing a die. The result is either heads or tails on the coin and a number from one to six on the die. The probability of a simple event is calculated by dividing the number of possible outcomes by the total number of outcomes. Therefore, the probability of obtaining heads on a coin is $\frac{1}{2}$, and the probability of rolling a 6 on a die is $\frac{1}{6}$. The probability of compound events is calculated using the basic idea of the probability of simple events. If the two events are independent, the probability of one outcome is equal to the product of the probabilities of each simple event. For example, the probability of obtaining heads on a coin and rolling a 6 is equal to:

$$\frac{1}{2} \times \frac{1}{6} = \frac{1}{12}$$

The probability of either A or B occurring is equal to the sum of the probabilities minus the probability that both A and B will occur. Therefore, the probability of obtaining either heads on a coin or rolling a 6 on a die is:

$$\frac{1}{2} + \frac{1}{6} - \frac{1}{12} = \frac{7}{12}$$

The two events aren't mutually exclusive because they can happen at the same time. If two events are mutually exclusive, and the probability of both events occurring at the same time is zero, the probability of event A or B occurring equals the sum of both probabilities. An example of calculating the probability of two mutually exclusive events is determining the probability of pulling a king or a queen from a deck of cards. The two events cannot occur at the same time.

Basic Algebra

Adding, Subtracting, Multiplying, and Factoring Linear Expressions

Algebraic expressions look similar to equations, but they do not include the equal sign. Algebraic expressions are comprised of numbers, variables, and mathematical operations. Some examples of algebraic expressions are:

$$8x + 7y - 12z$$

$$3a^2$$

$$5x^3 - 4y^4$$

Algebraic expressions consist of variables, numbers, and operations. A term of an expression is any combination of numbers and/or variables, and terms are separated by addition and subtraction. For example, the expression:

$$5x^2 - 3xy + 4y - 2$$

consists of 4 terms: $5x^2$, $-3xy$, $4y$, and -2. Note that each term includes it's given sign (+ or $-$). The variable part of a term is a letter that represents an unknown quantity. The coefficient of a term is the number by which the variable is multiplied. For the term $4y$, the variable is y, and the coefficient is 4. Terms are identified by the power (or exponent) of its variable.

A number without a variable is referred to as a constant. If the variable is to the first power (x^1 or simply x), it is referred to as a linear term. A term with a variable to the second power (x^2) is quadratic, and a term to the third power (x^3) is cubic. Consider the expression:

$$x^3 + 3x - 1$$

The constant is –1. The linear term is $3x$. There is no quadratic term. The cubic term is x^3.

An algebraic expression can also be classified by how many terms exist in the expression. Any like terms should be combined before classifying. A monomial is an expression consisting of only one term. Examples of monomials are: 17, $2x$, and $-5ab^2$. A binomial is an expression consisting of two terms separated by addition or subtraction. Examples include $2x - 4$ and $-3y^2 + 2y$. A trinomial consists of 3 terms. For example:

$$5x^2 - 2x + 1$$

is a trinomial.

Algebraic expressions and equations can be used to represent real-life situations and model the behavior of different variables. For example, $2x + 5$ could represent the cost to play games at an arcade. In this case, 5 represents the price of admission to the arcade, and 2 represents the cost of each game played. To calculate the total cost, use the number of games played for x, multiply it by 2, and add 5.

Adding and Subtracting Linear Algebraic Expressions

An algebraic expression is simplified by combining like terms. A term is a number, variable, or product of a number and variables separated by addition and subtraction. For the algebraic expression:

$$3x^2 - 4x + 5 - 5x^2 + x - 3$$

the terms are $3x^2$, $-4x$, 5, $-5x^2$, x, and -3. Like terms have the same variables raised to the same powers (exponents). The like terms for the previous example are $3x^2$ and $-5x^2$, $-4x$ and x, 5 and -3. To combine like terms, the coefficients (numerical factor of the term including sign) are added, and the variables and their powers are kept the same. Note that if a coefficient is not written, it is an implied coefficient of 1 ($x = 1x$). The previous example will simplify to:

$$-2x^2 - 3x + 2$$

When adding or subtracting algebraic expressions, each expression is written in parentheses. The negative sign is distributed when necessary, and like terms are combined. Consider the following:

add $2a + 5b - 2$ to $a - 2b + 8c - 4$

The sum is set as follows:

$$(a - 2b + 8c - 4) + (2a + 5b - 2)$$

In front of each set of parentheses is an implied positive one, which, when distributed, does not change any of the terms. Therefore, the parentheses are dropped and like terms are combined:

$$a - 2b + 8c - 4 + 2a + 5b - 2$$

$$3a + 3b + 8c - 6$$

Consider the following problem:

Subtract $2a + 5b - 2$ from $a - 2b + 8c - 4$

The difference is set as follows:

$$(a - 2b + 8c - 4) - (2a + 5b - 2)$$

The implied one in front of the first set of parentheses will not change those four terms. However, distributing the implied –1 in front of the second set of parentheses will change the sign of each of those three terms:

$$a - 2b + 8c - 4 - 2a - 5b + 2$$

Combining like terms yields the simplified expression:

$$-a - 7b + 8c - 2$$

Distributive Property

The **distributive property** states that multiplying a sum (or difference) by a number produces the same result as multiplying each value in the sum (or difference) by the number and adding (or subtracting) the products. Using mathematical symbols, the distributive property states:

$$a(b + c) = ab + ac$$

The expression $4(3 + 2)$ is simplified using the order of operations. Simplifying inside the parentheses first produces 4×5, which equals 20. The expression $4(3 + 2)$ can also be simplified using the distributive property:

$$4(3 + 2)$$

$$4 \times 3 + 4 \times 2$$

$$12 + 8$$

$$20$$

Consider the following example: $4(3x - 2)$. The expression cannot be simplified inside the parentheses because 3*x* and –2 are not like terms and therefore cannot be combined. However, the expression can be simplified by using the distributive property and multiplying each term inside of the parentheses by the term outside of the parentheses: $12x - 8$. The resulting equivalent expression contains no like terms, so it cannot be further simplified.

Consider the expression:

$$(3x + 2y + 1) - (5x - 3) + 2(3y + 4)$$

Again, there are no like terms, but the distributive property is used to simplify the expression. Note there is an implied one in front of the first set of parentheses and an implied –1 in front of the second set of parentheses. Distributing the 1, –1, and 2 produces:

$$1(3x) + 1(2y) + 1(1) - 1(5x) - 1(-3) + 2(3y) + 2(4)$$

$$3x + 2y + 1 - 5x + 3 + 6y + 8$$

This expression contains like terms that are combined to produce the simplified expression:

$$-2x + 8y + 12$$

Algebraic expressions are tested to be equivalent by choosing values for the variables and evaluating both expressions. For example, $4(3x - 2)$ and $12x - 8$ are tested by substituting 3 for the variable *x* and calculating to determine if equivalent values result.

Evaluating Algebraic Expressions

To evaluate the expression, the given values for the variables are substituted (or replaced), and the expression is simplified using the order of operations. Parentheses should be used when substituting. Consider the following: Evaluate $a - 2b + ab$ for $a = 3$ and $b = -1$. To evaluate, any variable *a* is replaced with 3 and any variable *b* with –1, producing:

$$(3) - 2(-1) + (3)(-1)$$

Next, the order of operations is used to calculate the value of the expression, which is 2.

Let's try two more.

Evaluate:

$$\frac{1}{2}x^2 - 3$$

$$x = 4$$

The first step is to substitute in 4 for *x* in the expression:

$$\frac{1}{2}(4)^2 - 3$$

Then, the order of operations is used to simplify.

The exponent comes first, $\frac{1}{2}(16) - 3$, then the multiplication 8 – 3, and then, after subtraction, the solution is 5.

Evaluate:

$$4|5 - x| + 2y$$

$$x = 4$$

$$y = -3$$

The first step is to substitute 4 in for *x* and –3 in for *y* in the expression:

$$4|5 - 4| + 2(-3)$$

Then, the absolute value expression is simplified, which is:

$$|5 - 4| = |1| = 1$$

The expression is:

$$4(1) + 2(-3)$$

which can be simplified using the order of operations.

First is the multiplication, $4 + (-6)$; then addition yields an answer of –2.

Creating Algebraic Expressions

A linear expression is a statement about an unknown quantity expressed in mathematical symbols. The statement "five times a number added to forty" can be expressed as $5x + 40$. A linear equation is a statement in which two expressions (at least one containing a variable) are equal to each other. The statement "five times a number added to forty is equal to ten" can be expressed as:

$$5x + 40 = 10$$

Real world scenarios can also be expressed mathematically. Suppose a job pays its employees \$300 per week and \$40 for each sale made. The weekly pay is represented by the expression $40x + 300$ where *x* is the number of sales made during the week.

Consider the following scenario: Bob had \$20 and Tom had \$4. After selling 4 ice cream cones to Bob, Tom has as much money as Bob. The cost of an ice cream cone is an unknown quantity and can be represented by a variable (*x*). The amount of money Bob has after his purchase is four times the cost of an ice cream cone subtracted from his original:

$$\$20 \rightarrow 20 - 4x$$

The amount of money Tom has after his sale is four times the cost of an ice cream cone added to his original:

$$\$4 \rightarrow 4x + 4$$

After the sale, the amount of money that Bob and Tom have is equal:

$$20 - 4x = 4x + 4$$

Adding, Subtracting, Multiplying, Dividing, and Factoring Polynomials

An expression of the form ax^n, where n is a non-negative integer, is called a **monomial** because it contains one term. A sum of monomials is called a **polynomial**. For example, $-4x^3 + x$ is a polynomial, while $5x^7$ is a monomial. A function equal to a polynomial is called a **polynomial function**.

The monomials in a polynomial are also called the **terms** of the polynomial.

The constants that precede the variables are called **coefficients**.

The highest value of the exponent of x in a polynomial is called the **degree** of the polynomial. So, $-4x^3 + x$ has a degree of 3, while:

$$-2x^5 + x^3 + 4x + 1$$

has a degree of 5. When multiplying polynomials, the degree of the result will be the sum of the degrees of the two polynomials being multiplied.

Addition and subtraction operations can be performed on polynomials with like terms. **Like terms** refers to terms that have the same variable and exponent. The two following polynomials can be added together by collecting like terms:

$$(x^2 + 3x - 4) + (4x^2 - 7x + 8)$$

The x^2 terms can be added as:

$$x^2 + 4x^2 = 5x^2$$

The x terms can be added as:

$$3x + -7x = -4x$$

and the constants can be added as $-4 + 8 = 4$.

The following expression is the result of the addition:

$$5x^2 - 4x + 4$$

Let's try another:

$$(-2x^5 + x^3 + 4x + 1) + (-4x^3 + x)$$

$$-2x^5 + (1 - 4)x^3 + (4 + 1)x + 1$$

$$-2x^5 - 3x^3 + 5x + 1$$

Likewise, subtraction of polynomials is performed by subtracting coefficients of like powers of x. So:

$$(-2x^5 + x^3 + 4x + 1) - (-4x^3 + x)$$

$$-2x^5 + (1 + 4)x^3 + (4 - 1)x + 1$$

$$-2x^5 + 5x^3 + 3x + 1$$

To multiply two polynomials, multiply each term of the first polynomial by each term of the second polynomial and add the results. For example:

$$(4x^2 + x)(-x^3 + x)$$

$$4x^2(-x^3) + 4x^2(x) + x(-x^3) + x(x)$$

$$-4x^5 + 4x^3 - x^4 + x^2$$

In the case where each polynomial has two terms, like in this example, some students find it helpful to remember this as multiplying the First terms, then the Outer terms, then the Inner terms, and finally the Last terms, with the mnemonic FOIL. For longer polynomials, the multiplication process is the same, but there will be, of course, more terms, and there is no common mnemonic to remember each combination.

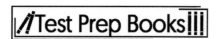

Factors for polynomials are similar to factors for integers—they are numbers, variables, or polynomials that, when multiplied together, give a product equal to the polynomial in question. One polynomial is a factor of a second polynomial if the second polynomial can be obtained from the first by multiplying by a third polynomial.

$$6x^6 + 13x^4 + 6x^2$$

can be obtained by multiplying together:

$$(3x^4 + 2x^2)(2x^2 + 3)$$

This means:

$$2x^2 + 3$$

and

$$3x^4 + 2x^2$$

are factors of:

$$6x^6 + 13x^4 + 6x^2$$

In general, finding the factors of a polynomial can be tricky. However, there are a few types of polynomials that can be factored in a straightforward way.

If a certain monomial is in each term of a polynomial, it can be factored out. There are several common forms polynomials take, which if you recognize, you can solve. The first example is a perfect square trinomial. To factor this polynomial, first expand the middle term of the expression:

$$x^2 + 2xy + y^2$$

$$x^2 + xy + xy + y^2$$

Factor out a common term in each half of the expression (in this case x from the left and y from the right):

$$x(x + y) + y(x + y)$$

Then the same can be done again, treating $(x + y)$ as the common factor:

$$(x + y)(x + y) = (x + y)^2$$

Therefore, the formula for this polynomial is:

$$x^2 + 2xy + y^2 = (x + y)^2$$

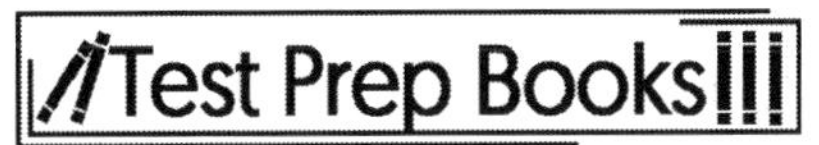

Next is another example of a perfect square trinomial. The process is the similar, but notice the difference in sign:

$$x^2 - 2xy + y^2$$

$$x^2 - xy - xy + y^2$$

Factor out the common term on each side:

$$x(x - y) - y(x - y)$$

Factoring out the common term again:

$$(x - y)(x - y) = (x - y)^2$$

Thus:

$$x^2 - 2xy + y^2 = (x - y)^2$$

The next is known as a difference of squares. This process is effectively the reverse of binomial multiplication:

$$x^2 - y^2$$

$$x^2 - xy + xy - y^2$$

$$x(x - y) + y(x - y)$$

$$(x + y)(x - y)$$

Therefore:

$$x^2 - y^2 = (x + y)(x - y)$$

The following two polynomials are known as the sum or difference of cubes. These are special polynomials that take the form of $x^3 + y^3$ or $x^3 - y^3$. The following formula factors the sum of cubes:

$$x^3 + y^3 = (x + y)(x^2 - xy + y^2)$$

Next is the difference of cubes, but note the change in sign. The formulas for both are similar, but the order of signs for factoring the sum or difference of cubes can be remembered by using the acronym SOAP, which stands for "same, opposite, always positive." The first sign is the same as the sign in the first expression, the second is opposite, and the third is always positive. The next formula factors the difference of cubes:

$$x^3 - y^3 = (x - y)(x^2 + xy + y^2)$$

The following two examples are expansions of cubed binomials. Similarly, these polynomials always follow a pattern:

$$x^3 + 3x^2y + 3xy^2 + y^3 = (x + y)^3$$

$$x^3 - 3x^2y + 3xy^2 - y^3 = (x - y)^3$$

These rules can be used in many combinations with one another. For example, the expression $3x^3 - 24$ has a common factor of 3, which becomes:

$$3(x^3 - 8)$$

A difference of cubes still remains which can then be factored out:

$$3(x - 2)(x^2 + 2x + 4)$$

There are no other terms to be pulled out, so this expression is completely factored.

When factoring polynomials, a good strategy is to multiply the factors to check the result. Let's try another example:

$$4x^3 + 16x^2$$

Both sides of the expression can be divided by 4, and both contain x^2, because $4x^3$ can be thought of as $4x^2(x)$, so the common term can simply be factored out:

$$4x^2(x + 4)$$

It sometimes can be necessary to rewrite the polynomial in some clever way before applying the above rules. Consider the problem of factoring $x^4 - 1$. This does not immediately look like any of the previous polynomials. However, it's possible to think of this polynomial as:

$$x^4 - 1 = (x^2)^2 - (1^2)^2$$

and now it can be treated as a difference of squares to simplify this:

$$(x^2)^2 - (1^2)^2$$

$$(x^2)^2 - x^2 1^2 + x^2 1^2 - (1^2)^2$$

$$x^2(x^2 - 1^2) + 1^2(x^2 - 1^2)$$

$$(x^2 + 1^2)(x^2 - 1^2)$$

$$(x^2 + 1)(x^2 - 1)$$

Creating Polynomials from Written Descriptions

Polynomials that represent mathematical or real-world problems can also be created from written descriptions, much like algebraic expressions. For example, polynomials might be created when working with formulas. Formulas are mathematical expressions that define the value of one quantity, given the value of one or more different quantities. Formulas look like equations because they contain variables, numbers, operators, and an equal sign. All formulas are equations, but not all equations are formulas. A formula must have more than one variable. For example, $2x + 7 = y$ is an equation and a formula (it relates the unknown quantities *x* and *y*). However, $2x + 7 = 3$ is an equation but not a formula (it only expresses the value of the unknown quantity *x*).

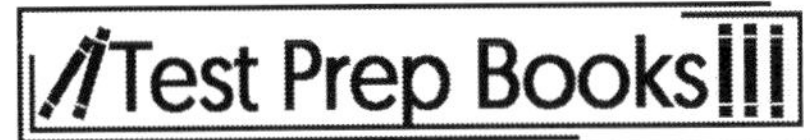

Formulas are typically written with one variable alone (or isolated) on one side of the equal sign. This variable can be thought of as the **subject** in that the formula is stating the value of the subject in terms of the relationship between the other variables. Consider the distance formula:

$$distance = rate \times time$$

or

$$d = rt$$

The value of the subject variable *d* (distance) is the product of the variable *r* and *t* (rate and time). Given the rate and time, the distance traveled can easily be determined by substituting the values into the formula and evaluating.

The formula $P = 2l + 2w$ expresses how to calculate the perimeter of a rectangle (*P*) given its length (*l*) and width (*w*). To find the perimeter of a rectangle with a length of 3ft and a width of 2ft, these values are substituted into the formula for *l* and *w*:

$$P = 2(3ft) + 2(2ft)$$

Following the order of operations, the perimeter is determined to be 10ft. When working with formulas such as these, including units is an important step.

Given a formula expressed in terms of one variable, the formula can be manipulated to express the relationship in terms of any other variable. In other words, the formula can be rearranged to change which variable is the *subject*. To solve for a variable of interest by manipulating a formula, the equation may be solved as if all other variables were numbers. The same steps for solving are followed, leaving operations in terms of the variables instead of calculating numerical values. For the formula $P = 2l + 2w$, the perimeter is the subject expressed in terms of the length and width. To write a formula to calculate the width of a rectangle, given its length and perimeter, the previous formula relating the three variables is solved for the variable *w*. If *P* and *l* were numerical values, this is a two-step linear equation solved by subtraction and division. To solve the equation $P = 2l + 2w$ for *w*, 2*l* is first subtracted from both sides:

$$P - 2l = 2w$$

Then both sides are divided by 2:

$$\frac{P - 2l}{2} = w$$

Test questions may involve creating a polynomial based on a formula. For example, using the perimeter of a rectangle formula, a problem may ask for the perimeter of a rectangle with a length of $2x + 12$ and a width of $x + 1$. Using the formula $P = 2l + 2w$, the perimeter would then be:

$$P = 2(2x + 12) + 2(x + 1)$$

This equals:

$$4x + 24 + 2x + 2 = 6x + 26$$

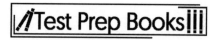

The area of the same rectangle, which uses the formula $A = l \times w$, would be:

$$A = (2x + 12)(x + 1)$$

$$2x^2 + 2x + 12x + 12$$

$$2x^2 + 14x + 12$$

Adding, Subtracting, Multiplying, Dividing Rational Expressions

A fraction, or ratio, wherein each part is a polynomial, defines **rational expressions**. Some examples include:

$$\frac{2x + 6}{x}$$

$$\frac{1}{x^2 - 4x + 8}$$

and

$$\frac{z^2}{x + 5}$$

Exponents on the variables are restricted to whole numbers, which means roots and negative exponents are not included in rational expressions.

Rational expressions can be transformed by factoring. For example, the expression:

$$\frac{x^2 - 5x + 6}{(x - 3)}$$

can be rewritten by factoring the numerator to obtain:

$$\frac{(x - 3)(x - 2)}{(x - 3)}$$

Therefore, the common binomial $(x - 3)$ can cancel so that the simplified expression is:

$$\frac{(x - 2)}{1} = (x - 2)$$

Additionally, other rational expressions can be rewritten to take on different forms. Some may be factorable in themselves, while others can be transformed through arithmetic operations. Rational expressions are closed under addition, subtraction, multiplication, and division by a nonzero expression. **Closed** means that if any one of these operations is performed on a rational expression, the result will still be a rational expression. The set of all real numbers is another example of a set closed under all four operations.

Adding and subtracting rational expressions is based on the same concepts as adding and subtracting simple fractions. For both concepts, the denominators must be the same for the operation to take place. For example, here are two rational expressions:

$$\frac{x^3 - 4}{(x - 3)} + \frac{x + 8}{(x - 3)}$$

Since the denominators are both $(x - 3)$, the numerators can be combined by collecting like terms to form:

$$\frac{x^3 + x + 4}{(x - 3)}$$

If the denominators are different, they need to be made common (the same) by using the **Least Common Denominator (LCD)**. Each denominator needs to be factored, and the LCD contains each factor that appears in any one denominator the greatest number of times it appears in any denominator. The original expressions need to be multiplied by a form of 1 such as $\frac{5}{5}$ or $\frac{x-2}{x-2}$, which will turn each denominator into the LCD. This process is like adding fractions with unlike denominators. It is also important when working with rational expressions to define what value of the variable makes the denominator zero. For this particular value, the expression is undefined.

Multiplication of rational expressions is performed like multiplication of fractions. The numerators are multiplied; then, the denominators are multiplied. The final fraction is then simplified. The expressions are simplified by factoring and cancelling out common terms. In the following example, the numerator of the second expression can be factored first to simplify the expression before multiplying:

$$\frac{x^2}{(x - 4)} \times \frac{x^2 - x - 12}{2}$$

$$\frac{x^2}{(x - 4)} \times \frac{(x - 4)(x + 3)}{2}$$

The $(x - 4)$ on the top and bottom cancel out:

$$\frac{x^2}{1} \times \frac{(x + 3)}{2}$$

Then multiplication is performed, resulting in:

$$\frac{x^3 + 3x^2}{2}$$

Dividing rational expressions is similar to the division of fractions, where division turns into multiplying by a reciprocal. Thus, the following expression can be rewritten as a multiplication problem:

$$\frac{x^2-3x+7}{x-4} \div \frac{x^2-5x+3}{x-4}$$

$$\frac{x^2-3x+7}{x-4} \times \frac{x-4}{x^2-5x+3}$$

The $x-4$ cancels out, leaving:

$$\frac{x^2-3x+7}{x^2-5x+3}$$

The final answers should always be completely simplified. If a function is composed of a rational expression, the zeros of the graph can be found from setting the polynomial in the numerator as equal to zero and solving. The values that make the denominator equal to zero will either exist on the graph as a **hole** or a **vertical asymptote**.

A **complex fraction** is a fraction in which the numerator and denominator are themselves fractions, of the form:

$$\frac{(\frac{a}{b})}{(\frac{c}{d})}$$

These can be simplified by following the usual rules for the order of operations, or by remembering that dividing one fraction by another is the same as multiplying by the reciprocal of the divisor. This means that any complex fraction can be rewritten using the following form:

$$\frac{(\frac{a}{b})}{(\frac{c}{d})} = \frac{a}{b} \times \frac{d}{c}$$

The following problem is an example of solving a complex fraction:

$$\frac{(\frac{5}{4})}{(\frac{3}{8})} = \frac{5}{4} \times \frac{8}{3} = \frac{40}{12} = \frac{10}{3}$$

Writing an Expression from a Written Description

When expressing a verbal or written statement mathematically, it is vital to understand words or phrases that can be represented with symbols. The following are examples:

Symbol	Phrase
+	Added to; increased by; sum of; more than
−	Decreased by; difference between; less than; take away
×	Multiplied by; 3(4,5...) times as large; product of
÷	Divided by; quotient of; half (third, etc.) of
=	Is; the same as; results in; as much as; equal to
x,t,n, etc.	A number; unknown quantity; value of; variable

Addition and subtraction are **inverse operations**. Adding a number and then subtracting the same number will cancel each other out, resulting in the original number, and vice versa. For example:

$$8 + 7 - 7 = 8$$

and

$$137 - 100 + 100 = 137$$

Similarly, multiplication and division are inverse operations. Therefore, multiplying by a number and then dividing by the same number results in the original number, and vice versa. For example:

$$8 \times 2 \div 2 = 8$$

and

$$12 \div 4 \times 4 = 12$$

Inverse operations are used to work backwards to solve problems. In the case that 7 and a number add to 18, the inverse operation of subtraction is used to find the unknown value ($18 - 7 = 11$). If a school's entire 4th grade was divided evenly into 3 classes each with 22 students, the inverse operation of multiplication is used to determine the total students in the grade ($22 \times 3 = 66$).

Recall that a rational expression is a fraction where the numerator and denominator are both polynomials.

Some examples of rational expressions include the following:

$$\frac{4x^3y^5}{3z^4}$$

$$\frac{4x^3 + 3x}{x^2}$$

and

$$\frac{x^2 + 7x + 10}{x + 2}$$

Since these refer to expressions and not equations, they can be simplified but not solved. Using the rules of exponents and roots, some rational expressions with monomials can be simplified. Other rational expressions such as the last example:

$$\frac{x^2 + 7x + 10}{x + 2}$$

take more steps to be simplified. First, the polynomial on top can be factored from:

$$x^2 + 7x + 10$$

into

$$(x + 5)(x + 2)$$

Then the common factors can be canceled, and the expression can be simplified to $(x + 5)$.

Consider this problem as an example of using rational expressions. Reggie wants to lay sod in his rectangular backyard. The length of the yard is given by the expression $4x + 2$, and the width is unknown. The area of the yard is $20x + 10$. Reggie needs to find the width of the yard. Knowing that the area of a rectangle is length multiplied by width, an expression can be written to find the width:

$$\frac{20x + 10}{4x + 2}$$

area divided by length. Simplifying this expression by factoring out 10 on the top and 2 on the bottom leads to this expression:

$$\frac{10(2x + 1)}{2(2x + 1)}$$

By cancelling out the $2x + 1$, that results in $\frac{10}{2} = 5$. The width of the yard is found to be 5 by simplifying a rational expression.

Using Linear Equations to Solve Real-World Problems

Linear relationships describe the way two quantities change with respect to each other. The relationship is defined as linear because a line is produced if all the sets of corresponding values are graphed on a

coordinate grid. When expressing the linear relationship as an equation, the equation is often written in the form $y = mx + b$ (slope-intercept form) where *m* and *b* are numerical values and *x* and *y* are variables (for example, $y = 5x + 10$). Given a linear equation and the value of either variable (*x* or *y*), the value of the other variable can be determined.

Imagine the following problem: The sum of a number and 5 is equal to –8 times the number.

To find this unknown number, a simple equation can be written to represent the problem. Key words such as difference, equal, and times are used to form the following equation with one variable:

$$n + 5 = -8n$$

When solving for n, opposite operations are used. First, n is subtracted from $-8n$ across the equals sign, resulting in $5 = -9n$. Then, –9 is divided on both sides, leaving $n = -\frac{5}{9}$. This solution can be graphed on the number line with a dot as shown below:

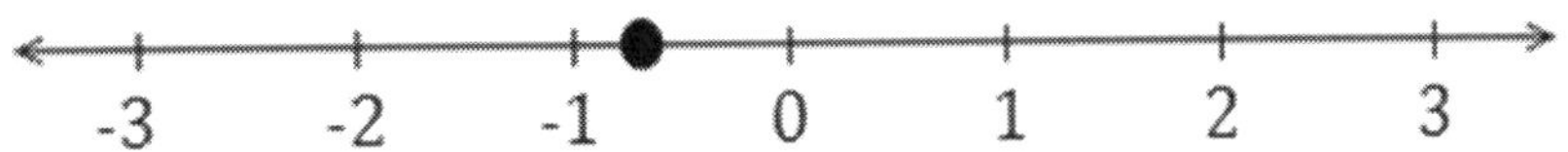

Suppose a teacher is grading a test containing 20 questions with 5 points given for each correct answer, adding a curve of 10 points to each test. This linear relationship can be expressed as the equation:

$$y = 5x + 10$$

where *x* represents the number of correct answers, and *y* represents the test score. To determine the score of a test with a given number of correct answers, the number of correct answers is substituted into the equation for *x* and evaluated. For example, for 10 correct answers, 10 is substituted for *x*:

$$y = 5(10) + 10 \rightarrow y = 60$$

Therefore, 10 correct answers will result in a score of 60. The number of correct answers needed to obtain a certain score can also be determined. To determine the number of correct answers needed to score a 90, 90 is substituted for *y* in the equation (*y* represents the test score) and solved:

$$90 = 5x + 10 \rightarrow 80 = 5x \rightarrow 16 = x$$

Therefore, 16 correct answers are needed to score a 90.

Linear relationships may be represented by a table of 2 corresponding values. Certain tables may determine the relationship between the values and predict other corresponding sets. Consider the table below, which displays the money in a checking account that charges a monthly fee:

Month	0	1	2	3	4
Balance	\$210	\$195	\$180	\$165	\$150

An examination of the values reveals that the account loses \$15 every month (the month increases by one and the balance decreases by 15). This information can be used to predict future values. To determine what the value will be in month 6, the pattern can be continued, and it can be concluded that the balance will be \$120. To determine which month the balance will be \$0, \$210 is divided by \$15 (since the balance decreases \$15 every month), resulting in month 14.

Solving a System of Two Linear Equations

A **system of equations** is a group of equations that have the same variables or unknowns. These equations can be linear, but they are not always so. Finding a solution to a system of equations means finding the values of the variables that satisfy each equation. For a linear system of two equations and two variables, there could be a single solution, no solution, or infinitely many solutions.

A single solution occurs when there is one value for x and y that satisfies the system. This would be shown on the graph where the lines cross at exactly one point. When there is no solution, the lines are parallel and do not ever cross. With infinitely many solutions, the equations may look different, but they are the same line. One equation will be a multiple of the other, and on the graph, they lie on top of each other.

The process of elimination can be used to solve a system of equations. For example, the following equations make up a system:

$$x + 3y = 10 \text{ and } 2x - 5y = 9$$

Immediately adding these equations does not eliminate a variable, but it is possible to change the first equation by multiplying the whole equation by -2. This changes the first equation to

$$-2x - 6y = -20$$

The equations can be then added to obtain $-11y = -11$. Solving for y yields $y = 1$. To find the rest of the solution, 1 can be substituted in for y in either original equation to find the value of $x = 7$. The solution to the system is (7, 1) because it makes both equations true, and it is the point in which the lines intersect. If the system is **dependent**—having infinitely many solutions—then both variables will cancel out when the elimination method is used, resulting in an equation that is true for many values of *x* and *y*. Since the system is dependent, both equations can be simplified to the same equation or line.

A system can also be solved using **substitution.** This involves solving one equation for a variable and then plugging that solved equation into the other equation in the system. For example:

$$x - y = -2$$

and

$$3x + 2y = 9$$

can be solved using substitution. The first equation can be solved for x, where $x = -2 + y$. Then it can be plugged into the other equation:

$$3(-2 + y) + 2y = 9$$

Solving for y yields:

$$-6 + 3y + 2y = 9$$

That shows that $y = 3$. If $y = 3$, then $x = 1$.

This solution can be checked by plugging in these values for the variables in each equation to see if it makes a true statement.

Finally, a solution to a system of equations can be found graphically. The solution to a linear system is the point or points where the lines cross. The values of x and y represent the coordinates (x, y) where the lines intersect. Using the same system of equations as above, they can be solved for y to put them in slope-intercept form, $y = mx + b$.

These equations become:

$$y = x + 2$$

and

$$y = -\frac{3}{2}x + 4.5$$

The slope is the coefficient of x, and the y-intercept is the constant value.

This system with the solution is shown below:

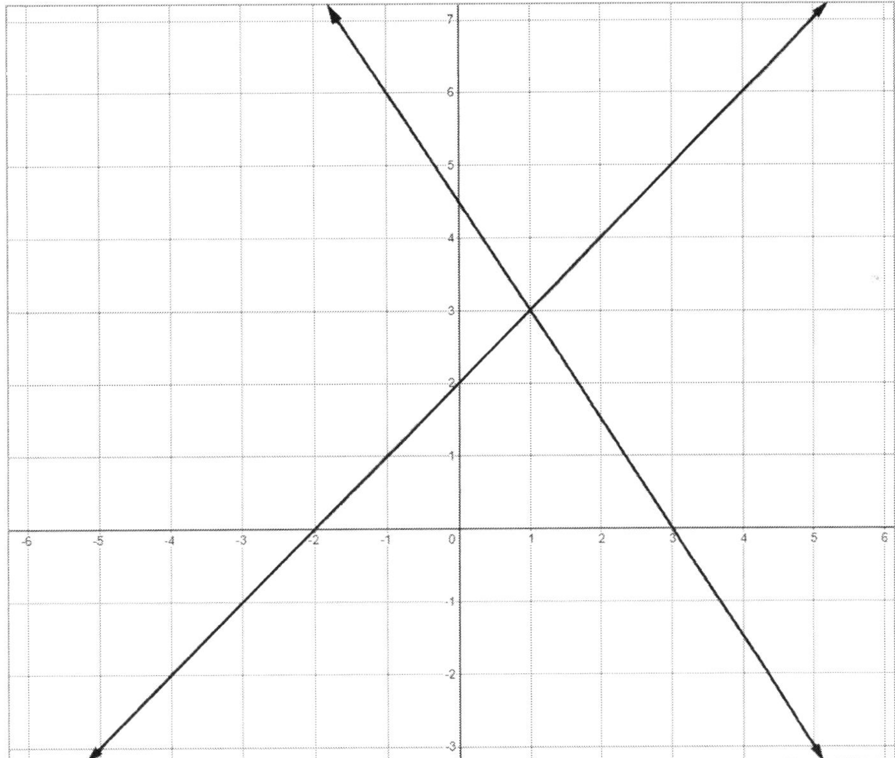

If the lines intersect, the point of intersection is the solution to the system. Every point on a line represents an ordered pair that makes its equation true. The ordered pair represented by this point of intersection lies on both lines and therefore makes both equations true. This ordered pair should be checked by substituting its values into both of the original equations of the system. Note that given a system of equations and an ordered pair, the ordered pair can be determined to be a solution or not by checking it in both equations.

If, when graphed, the lines representing the equations of a system do not intersect, then the two lines are parallel to each other or they are the same exact line. Parallel lines extend in the same direction without ever meeting. A system consisting of parallel lines has no solution. If the equations for a system represent the same exact line, then every point on the line is a solution to the system. In this case, there would be an infinite number of solutions. A system consisting of intersecting lines is referred to as independent; a system consisting of parallel lines is referred to as inconsistent; and a system consisting of coinciding lines is referred to as dependent.

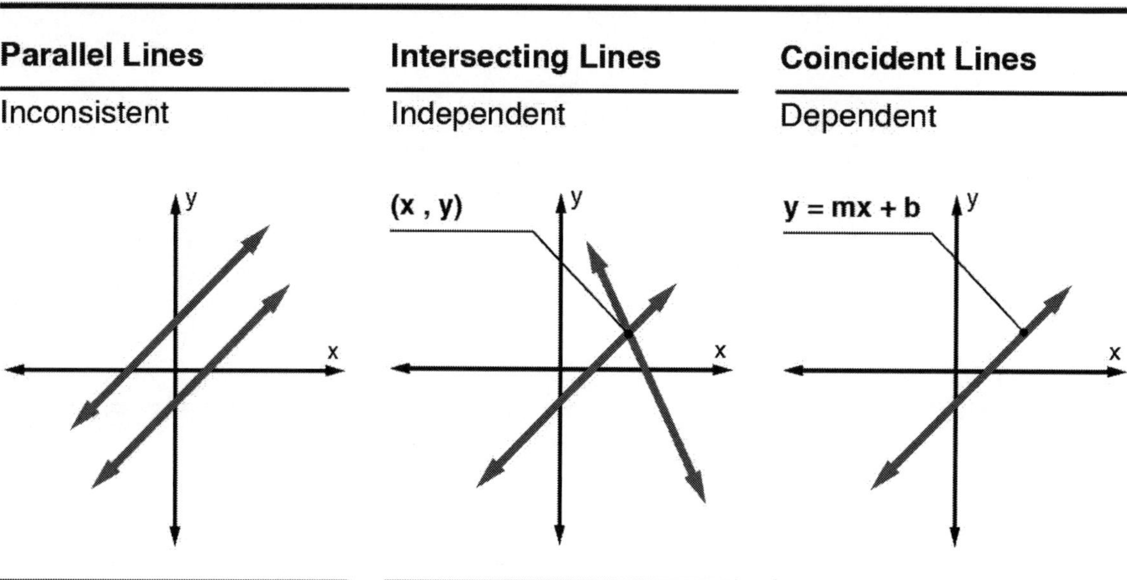

Matrices can also be used to solve systems of linear equations. Specifically, for systems, the coefficients of the linear equations in standard form are the entries in the matrix. Using the same system of linear equations as above, $x - y = -2$ and $3x + 2y = 9$, the matrix to represent the system is:

$$\begin{bmatrix} 1 & -1 \\ 3 & 2 \end{bmatrix} \begin{bmatrix} x \\ y \end{bmatrix} = \begin{bmatrix} -2 \\ 9 \end{bmatrix}$$

To solve this system using matrices, the inverse matrix must be found. For a general 2×2 matrix:

$$\begin{bmatrix} a & b \\ c & d \end{bmatrix}$$

The inverse matrix is found by the expression:

$$\frac{1}{ad - bc} \begin{bmatrix} d & -b \\ -c & a \end{bmatrix}$$

The inverse matrix for the system given above is:

$$\frac{1}{2 - -3}\begin{bmatrix} 2 & 1 \\ -3 & 1 \end{bmatrix} = \frac{1}{5}\begin{bmatrix} 2 & 1 \\ -3 & 1 \end{bmatrix}$$

The next step in solving is to multiply this identity matrix by the system matrix above. This is given by the following equation:

$$\frac{1}{5}\begin{bmatrix} 2 & 1 \\ -3 & 1 \end{bmatrix}\begin{bmatrix} 1 & -1 \\ 3 & 2 \end{bmatrix}\begin{bmatrix} x \\ y \end{bmatrix} = \begin{bmatrix} 2 & 1 \\ -3 & 1 \end{bmatrix}\begin{bmatrix} -2 \\ 9 \end{bmatrix}\frac{1}{5}$$

which simplifies to

$$\frac{1}{5}\begin{bmatrix} 5 & 0 \\ 0 & 5 \end{bmatrix}\begin{bmatrix} x \\ y \end{bmatrix} = \frac{1}{5}\begin{bmatrix} 5 \\ 15 \end{bmatrix}$$

Solving for the solution matrix, the answer is:

$$\begin{bmatrix} 1 & 0 \\ 0 & 1 \end{bmatrix}\begin{bmatrix} x \\ y \end{bmatrix} = \begin{bmatrix} 1 \\ 3 \end{bmatrix}$$

Since the first matrix is the identity matrix, the solution is $x = 1$ and $y = 3$.

Finding solutions to systems of equations is essentially finding what values of the variables make both equations true. It is finding the input value that yields the same output value in both equations. For functions $g(x)$ and $f(x)$, the equation $g(x) = f(x)$ means the output values are being set equal to each other. Solving for the value of x means finding the x-coordinate that gives the same output in both functions.

For example:

$$f(x) = x + 2$$

and

$$g(x) = -3x + 10$$

is a system of equations. Setting $f(x) = g(x)$ yields the equation:

$$x + 2 = -3x + 10$$

Solving for x, gives the x-coordinate $x = 2$ where the two lines cross. This value can also be found by using a table or a graph. On a table, both equations can be given the same inputs, and the outputs can be recorded to find the point(s) where the lines cross. Any method of solving finds the same solution, but some methods are more appropriate for some systems of equations than others.

Solving Inequalities and Graphing the Answer on a Number Line

Linear inequalities and linear equations are both comparisons of two algebraic expressions. However, unlike equations in which the expressions are equal to each other, linear inequalities compare expressions that are unequal. Linear equations typically have one value for the variable that makes the statement true. Linear inequalities generally have an infinite number of values that make the statement true.

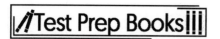

If a problem were to say, "The sum of a number and 5 is greater than –8 times the number," then an inequality would be used instead of an equation. Using key words again, *greater than* is represented by the symbol >. The inequality"

$$n + 5 > -8n$$

can be solved using the same techniques, resulting in:

$$n < -\frac{5}{9}$$

The only time solving an inequality differs from solving an equation is when a negative number is either multiplied by or divided by each side of the inequality. The sign must be switched in this case. For this example, the graph of the solution changes to the following graph because the solution represents all real numbers less than $-\frac{5}{9}$. Not included in this solution is $-\frac{5}{9}$ because it is a *less than* symbol, not *equal to*.

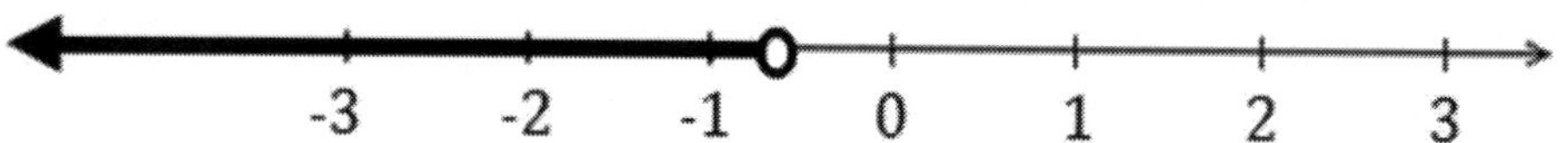

When solving a linear inequality, the solution is the set of all numbers that makes the statement true. The inequality $x + 2 \geq 6$ has a solution set of 4 and every number greater than 4 (4.0001, 5, 12, 107, etc.). Adding 2 to 4 or any number greater than 4 would result in a value that is greater than or equal to 6. Therefore, $x \geq 4$ would be the solution set.

Solution sets for linear inequalities often will be displayed using a number line. If a value is included in the set ($\geq$ or $\leq$), there is a shaded dot placed on that value and an arrow extending in the direction of the solutions. For a variable $>$ or $\geq$ a number, the arrow would point right on the number line (the direction where the numbers increase); and if a variable is $<$ or $\leq$ a number, the arrow would point left (where the numbers decrease). If the value is not included in the set ($>$ or $<$), an open circle on that value would be used with an arrow in the appropriate direction.

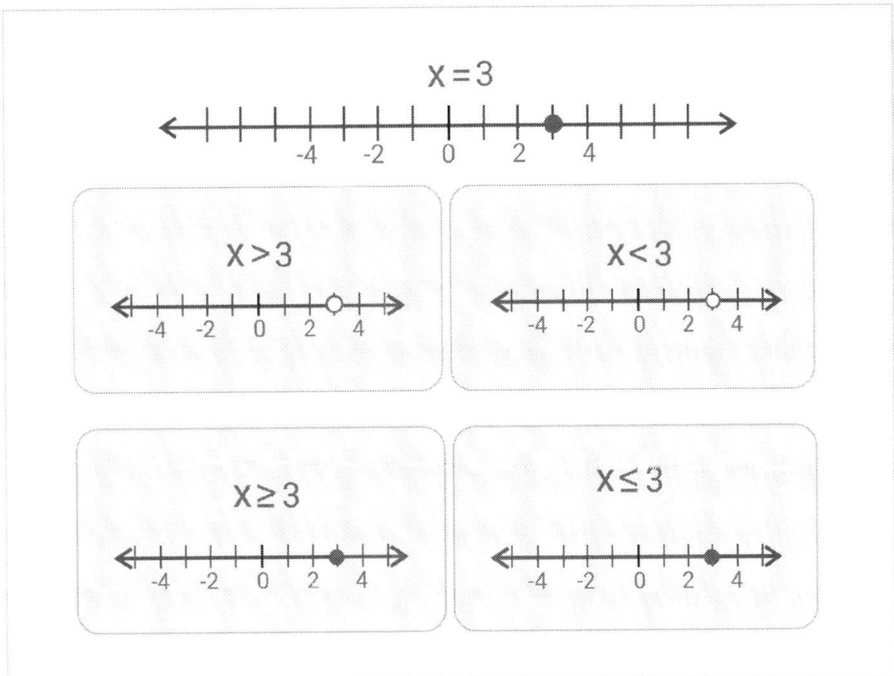

Students may be asked to write a linear inequality given a graph of its solution set. To do so, they should identify whether the value is included (shaded dot or open circle) and the direction in which the arrow is pointing.

In order to algebraically solve a linear inequality, the same steps should be followed as in solving a linear equation. The inequality symbol stays the same for all operations EXCEPT when multiplying or dividing by a negative number. If multiplying or dividing by a negative number while solving an inequality, the relationship reverses (the sign flips). Multiplying or dividing by a positive does not change the relationship, so the sign stays the same. In other words, $>$ switches to $<$ and vice versa. An example is shown below:

Solve $-2(x + 4) \leq 22$ for the value of x.

First, distribute –2 to the binomial by multiplying:

$$-2x - 8 \leq 22$$

Next, add 8 to both sides to isolate the variable:

$$-2x \leq 30$$

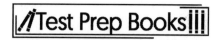

Divide both sides by −2 to solve for x:

$$x \geq -15$$

With a single equation in two variables, the solutions are limited only by the situation the equation represents. When two equations or inequalities are used, more constraints are added. For example, in a system of linear equations, there is often—although not always—only one answer. The point of intersection of two lines is the solution. For a system of inequalities, there are infinitely many answers.

The intersection of two solution sets gives the solution set of the system of inequalities. In the following graph, the darker shaded region is where two inequalities overlap. Any set of x and y found in that region satisfies both inequalities. The line with the positive slope is solid, meaning the values on that line are included in the solution.

The line with the negative slope is dotted, so the coordinates on that line are not included.

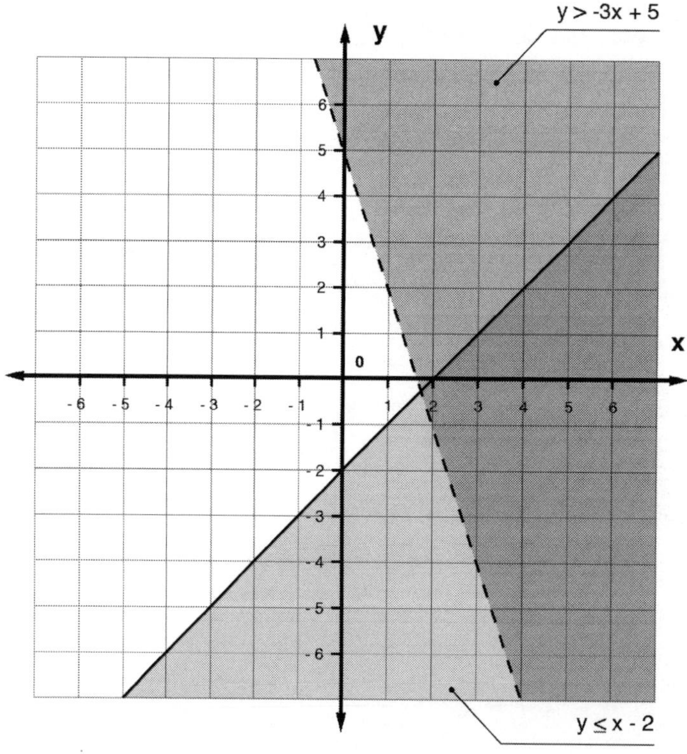

Quadratic Equations with One Variable

A **quadratic equation** can be written in the form:

$$y = ax^2 + bx + c$$

The u-shaped graph of a quadratic equation is called a **parabola**. The graph can either open up or open down (upside down u). The graph is symmetric about a vertical line, called the **axis of symmetry**. Corresponding points on the parabola are directly across from each other (same y-value) and are the

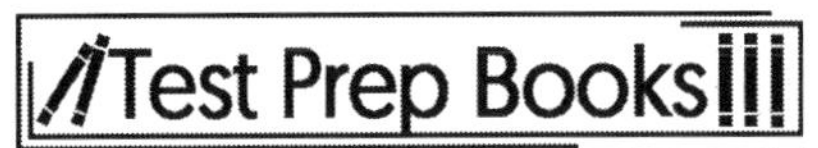

same distance from the axis of symmetry (on either side). The axis of symmetry intersects the parabola at its **vertex**. The y-value of the vertex represents the minimum or maximum value of the function. If the graph opens up, the value of a in its equation is positive, and the vertex represents the minimum of the function. If the graph opens down, the value of a in its equation is negative, and the vertex represents the maximum of the function.

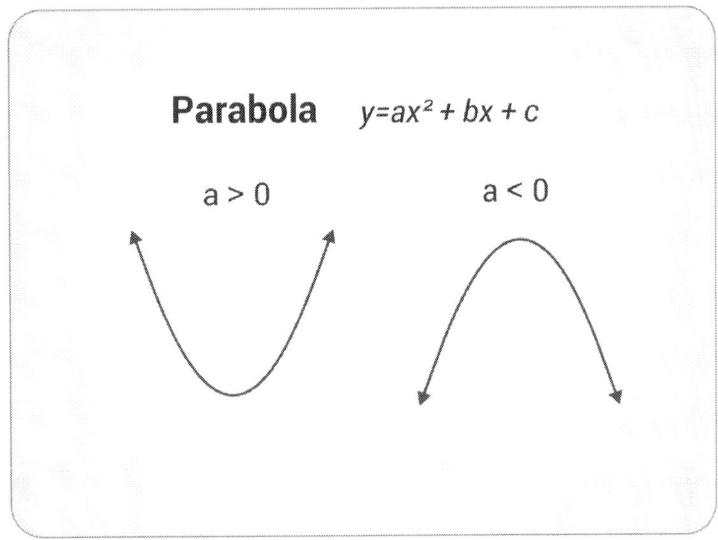

For a quadratic equation where the value of a is positive, as the inputs increase, the outputs increase until a certain value (maximum of the function) is reached. As inputs increase past the value that corresponds with the maximum output, the relationship reverses, and the outputs decrease. For a quadratic equation where a is negative, as the inputs increase, the outputs (1) decrease, (2) reach a maximum, and (3) then increase.

Consider a ball thrown straight up into the air. As time passes, the height of the ball increases until it reaches its maximum height. After reaching the maximum height, as time increases, the height of the ball decreases (it is falling toward the ground). This relationship can be expressed as a quadratic equation where time is the input (x), and the height of the ball is the output (y).

Equations with one variable (linear equations) can be solved using the addition principle and multiplication principle. If $a = b$, then $a + c = b + c$, and $ac = bc$. Given the equation"

$$2x - 3 = 5x + 7$$

the first step is to combine the variable terms and the constant terms. Using the principles, expressions can be added and subtracted onto and off both sides of the equals sign, so the equation turns into $-10 = 3x$. Dividing by 3 on both sides through the multiplication principle with $c = \frac{1}{3}$ results in the final answer of $x = \frac{-10}{3}$.

However, this same process cannot be used to solve nonlinear equations, including quadratic equations. Quadratic equations have a higher degree than linear ones (2 versus 1) and are not solved by simply using opposite operations. When an equation has a degree of 2, completing the square is an option. For example, the quadratic equation:

$$x^2 - 6x + 2 = 0$$

can be rewritten by completing the square. The goal of completing the square is to get the equation into the form:

$$(x - p)^2 = q$$

Using the example, the constant term 2 first needs to be moved over to the opposite side by subtracting. Then, the square can be completed by adding 9 to both sides, which is the square of half of the coefficient of the middle term $-6x$. The current equation is:

$$x^2 - 6x + 9 = 7$$

The left side can be factored into a square of a binomial, resulting in:

$$(x - 3)^2 = 7$$

To solve for x, the square root of both sides should be taken, resulting in:

$$(x - 3) = \pm\sqrt{7}$$

and

$$x = 3 \pm \sqrt{7}$$

Other ways of solving quadratic equations include graphing, factoring, and using the quadratic formula. The equation:

$$y = x^2 - 4x + 3$$

can be graphed on the coordinate plane, and the solutions can be observed where it crosses the x-axis. The graph will be a parabola that opens up with two solutions at 1 and 3.

If quadratic equations take the form $ax^2 - b = 0$, then the equation can be solved by adding *b* to both sides and dividing by *a* to get:

$$x^2 = \frac{b}{a} \text{ or } x = \pm\sqrt{\frac{b}{a}}$$

Note that this is actually two separate solutions, unless *b* happens to be 0.

If a quadratic equation has no constant—so that it takes the form:

$$ax^2 + bx = 0$$

then the *x* can be factored out to get

$$x(ax + b) = 0$$

Then, the solutions are $x = 0$, together with the solutions to

$$ax + b = 0$$

Both factors x and $(ax + b)$ can be set equal to zero to solve for x because one of those values must be zero for their product to equal zero. For an equation $ab = 0$ to be true, either $a = 0$, or $b = 0$.

A given quadratic equation:

$$x^2 + bx + c$$

can be factored into:

$$(x + A)(x + B)$$

where $A + B = b$, and $AB = c$. Finding the values of A and B can take time, but such a pair of numbers can be found by guessing and checking. Looking at the positive and negative factors for *c* offers a good starting point.

For example, in:

$$x^2 - 5x + 6$$

the factors of 6 are 1, 2, and 3. Now,$(-2)(-3) = 6$, and $-2 - 3 = -5$. In general, however, this may not work, in which case another approach may need to be used.

A quadratic equation of the form:

$$x^2 + 2xb + b^2 = 0$$

can be factored into $(x + b)^2 = 0$. Similarly

$$x^2 - 2xy + y^2 = 0$$

factors into $(x - y)^2 = 0$.

The first method of completing the square can be used in finding the second method, the quadratic formula. It can be used to solve any quadratic equation. This formula may be the longest method for solving quadratic equations and is commonly used as a last resort after other methods are ruled out.

It can be helpful in memorizing the formula to see where it comes from, so here are the steps involved.

The most general form for a quadratic equation is:

$$ax^2 + bx + c = 0$$

First, dividing both sides by *a* leaves us with:

$$x^2 + \frac{b}{a}x + \frac{c}{a} = 0$$

To complete the square on the left-hand side, c/a can be subtracted on both sides to get:

$$x^2 + \frac{b}{a}x = -\frac{c}{a}$$

$(\frac{b}{2a})^2$ is then added to both sides.

This gives:

$$x^2 + \frac{b}{a}x + (\frac{b}{2a})^2 = (\frac{b}{2a})^2 - \frac{c}{a}$$

The left can now be factored and the right-hand side simplified to give:

$$(x + \frac{b}{2a})^2 = \frac{b^2 - 4ac}{4a}$$

Taking the square roots gives:

$$x + \frac{b}{2a} = \pm\frac{\sqrt{b^2 - 4ac}}{2a}$$

Solving for x yields the quadratic formula:

$$x = \frac{-b \pm \sqrt{b^2 - 4ac}}{2a}$$

It isn't necessary to remember how to get this formula but memorizing the formula itself is the goal.

If an equation involves taking a root, then the first step is to move the root to one side of the equation and everything else to the other side. That way, both sides can be raised to the index of the radical in order to remove it, and solving the equation can continue.

Graphs and Functions

Locating Points and Graphing Equations

The coordinate plane, sometimes referred to as the Cartesian plane, is a two-dimensional surface consisting of a horizontal and a vertical number line. The horizontal number line is referred to as the *x*-axis, and the vertical number line is referred to as the *y*-axis. The *x*-axis and *y*-axis intersect (or cross) at a point called the origin. At the origin, the value of the *x*-axis is zero, and the value of the *y*-axis is zero. The coordinate plane identifies the exact location of a point that is plotted on the two-dimensional surface. Like a map, the location of all points on the plane are in relation to the origin. Along the *x*-axis (horizontal line), numbers to the right of the origin are positive and increasing in value (1,2,3, ...) and to the left of the origin numbers are negative and decreasing in value (–1,–2,–3, ...). Along the *y*-axis (vertical line), numbers above the origin are positive and increasing in value and numbers below the origin are negative and decreasing in value.

The *x*- and *y*-axis divide the coordinate plane into four sections. These sections are referred to as quadrant one, quadrant two, quadrant three, and quadrant four, and are often written with Roman numerals I, II, III, and IV.

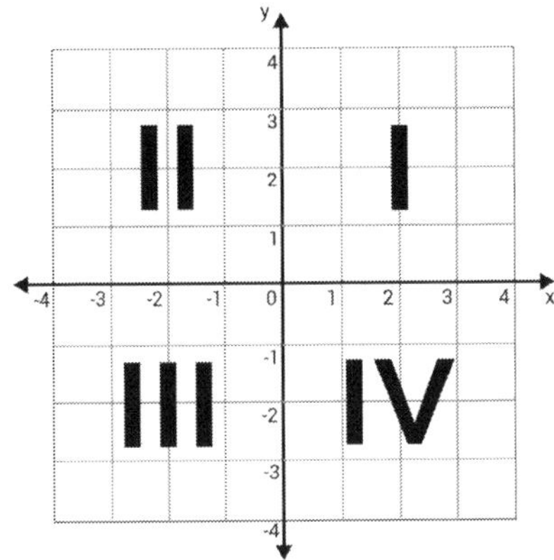

The upper right section is Quadrant I and consists of points with positive *x*-values and positive *y*-values. The upper left section is Quadrant II and consists of points with negative *x*-values and positive *y*-values. The bottom left section is Quadrant III and consists of points with negative *x*-values and negative *y*-values. The bottom right section is Quadrant IV and consists of points with positive *x*-values and negative *y*-values.

Graphing in the Coordinate Plane

The coordinate plane represents a representation of real-world space, and any point within the plane can be defined by a set of **coordinates** (x, y). The coordinates consist of two numbers, x and y, which represent a position on each number line. The coordinates can also be referred to as an **ordered pair,** and $(0,0)$ is the ordered pair known as the **vertex**, or the origin, the point in which the axes intersect.

Here is an example of the coordinate plane with a point plotted:

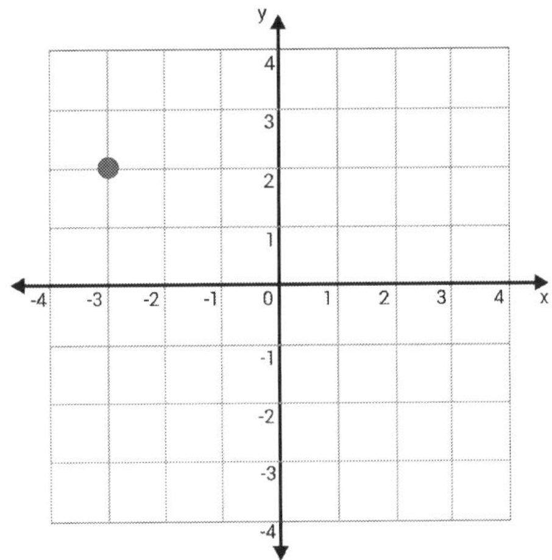

In order to plot a point on the coordinate plane, each coordinate must be considered individually. The value of x represents how many units away from the vertex the point lies on the x-axis. The value of y represents the number of units away from the vertex that the point lies on the y-axis.

For example, given the ordered pair (5, 4), the x-coordinate, 5, is the distance from the origin along the x-axis, and the y-coordinate, 4, is the distance from the origin along the y-axis. This is determined by counting 5 units to the right from (0, 0) along the x-axis and then counting 4 units up from that point, to reach the point where $x = 5$ and $y = 4$. In order to graph the single point, the point should be marked there with a dot and labeled as (5, 4). Every point on the plane has its own ordered pair.

Graphing on the Coordinate Plane Using Mathematical Problems, Tables, and Patterns

Data can be recorded using a coordinate plane. Graphs are utilized frequently in real-world applications and can be seen in many facets of everyday life. A relationship can exist between the *x*- and *y*-coordinates that are plotted on a graph, and those values can represent a set of data that can be listed in a table. Going back and forth between the table and the graph is an important concept and defining the relationship between the variables is the key that links the data to a real-life application.

For example, temperature increases during a summer day. The *x*-coordinate can be used to represent hours in the day, and the *y*-coordinate can be used to represent the temperature in degrees. The graph would show the temperature at each hour of the day. Time is almost always plotted on the *x*-axis, and utilizing different units on each axis, if necessary, is important. Labeling the axes with units is also important.

Within the first quadrant of the coordinate plane, both the x and y values are positive. Most real-world problems can be plotted in this quadrant because most real-world quantities, such as time and distance, are positive. Consider the following table of values:

X	Y
1	2
2	4
3	6
4	8

Each row gives a coordinate pair. For example, the first row gives the coordinates (1,2). Each *x*-value tells you how far to move from the origin, the point (0,0), to the right, and each *y*-value tells you how far to move up from the origin.

Here is the graph of the points listed above in the table in addition to the origin:

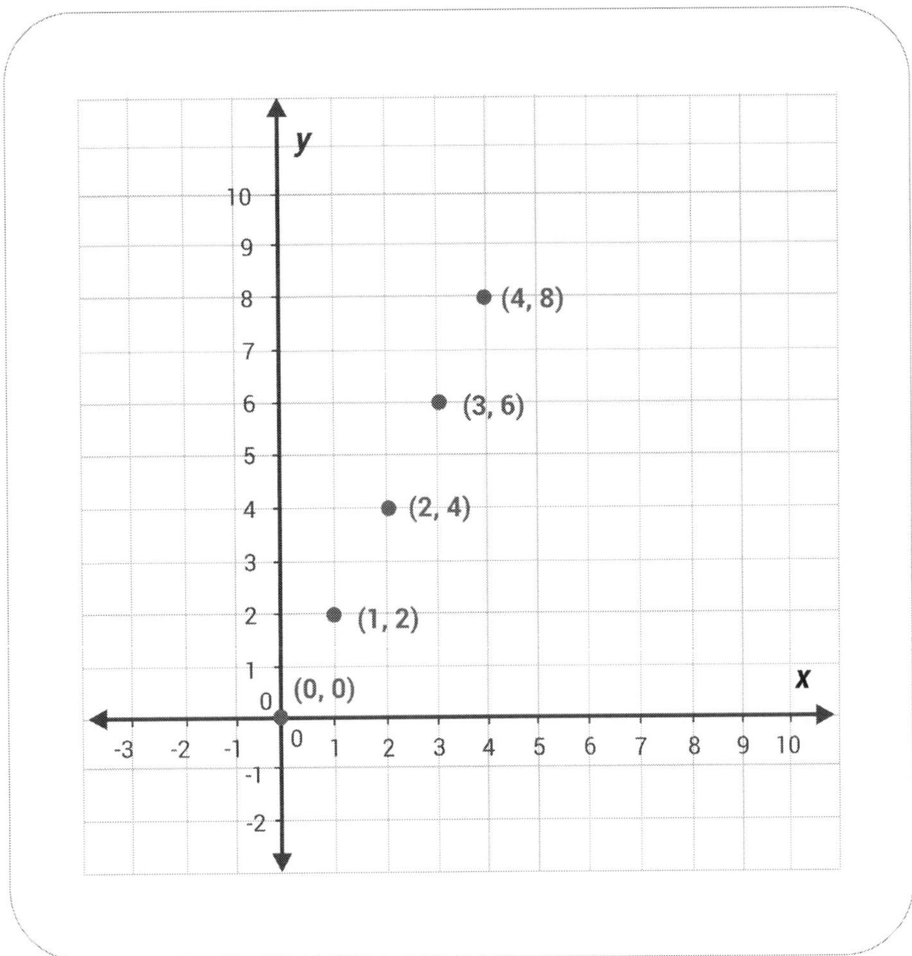

Notice that each *y*-value is found by doubling the *x*-value that forms the other portion of its coordinate pair.

Determining the Slope of a Line from a Graph, Equation, or Table

Rate of change for any line calculates the steepness of the line over a given interval. Rate of change is also known as the **slope** or rise/run. The slope of a linear function is given by the change in *y* divided by the change in *x*. So, the formula looks like this:

$$slope = \frac{y_2 - y_1}{x_2 - x_1}$$

In the graph below, two points are plotted. The first has the coordinates of (0, 1), and the second point is (2, 3). Remember that the x coordinate is always placed first in coordinate pairs. Work from left to right when identifying coordinates. Thus, the point on the left is point 1 (0, 1), and the point on the right is point 2 (2, 3).

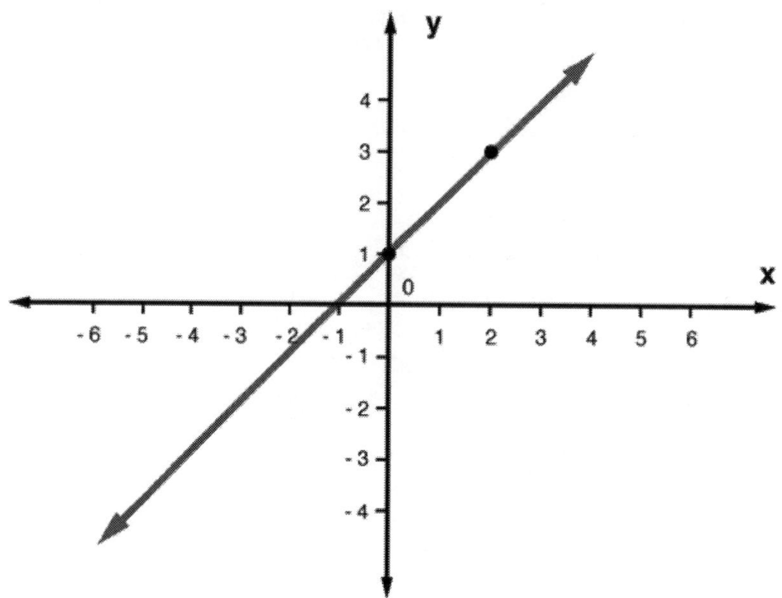

Now we need to just plug those numbers into the equation:

$$slope = \frac{3-1}{2-0}$$

$$slope = \frac{2}{2}$$

$$slope = 1$$

This means that for every increase of 1 for x, y also increased by 1. You can see this in the line. When x equaled 0, y equaled 1, and when x was increased to 1, y equaled 2.

Slope can be thought of as determining the rise over run:

$$slope = \frac{rise}{run}$$

The rise being the change vertically on the y axis and the run being the change horizontally on the x axis.

Proportional Relationships for Equations and Graphs

The rate of change for a linear function is constant and can be determined based on a few representations. One method is to place the equation in slope-intercept form: $y = mx + b$. Thus, m is the slope, and b is the y-intercept. In the graph below, the equation is $y = x + 1$, where the slope is 1 and the y-intercept is 1. For every vertical change of 1 unit, there is a horizontal change of 1 unit.

The x-intercept is -1, which is the point where the line crosses the x-axis:

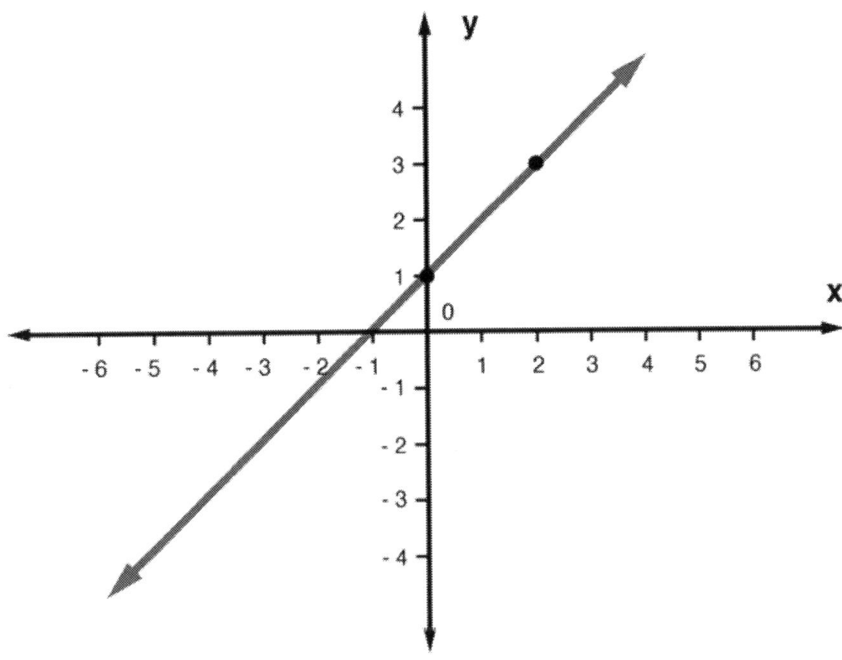

Let's look at an example of a proportional, or linear relationship, seen in the real world.

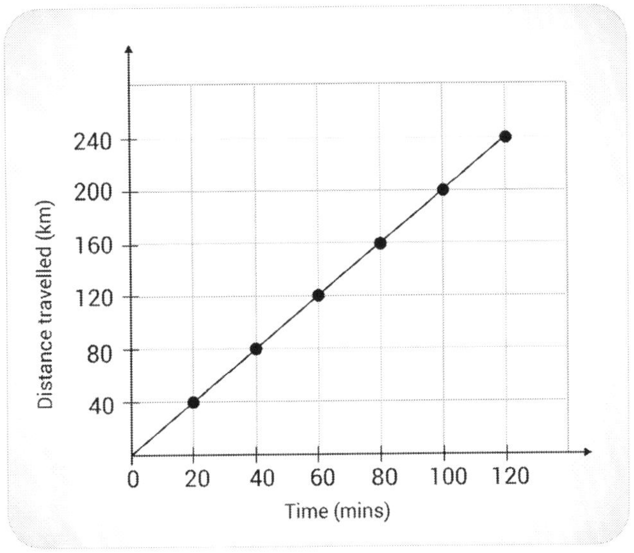

The graph above represents the relationship between distance traveled and time. To find the distance traveled in 80 minutes, the mark for 80 minutes is located at the bottom of the graph. By following this mark directly up on the graph, the corresponding point for 80 minutes is directly across from the 160 kilometer mark. This information indicates that the distance travelled in 80 minutes is 160 kilometers. To predict information not displayed on the graph, the way in which the variables change with respect to one another is determined. In this case, distance increases by 40 kilometers as time increases by 20 minutes. This information can be used to continue the data in the graph or convert the values to a table.

Let's try another example. Jim owns a car wash and charges $40 per car. The rent for the facility is $350 per month. An equation can be written to relate the number of cars Jim cleans to the money he makes per month. Let x represent the number of cars and y represent the profit Jim makes each month from the car wash. The equation $y = 40x - 350$ can be used to show Jim's profit or loss. Since this equation has two variables, the coordinate plane can be used to show the relationship and predict profit or loss for Jim. The following graph shows that Jim must wash at least nine cars to pay the rent, where $x = 9$. Anything nine cars and above yield a profit shown in the value on the y-axis.

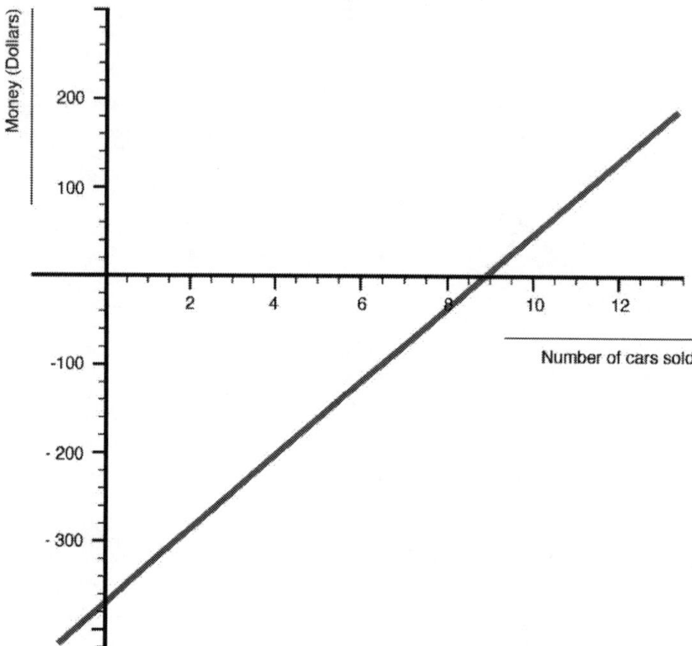

Formulas with two variables are equations used to represent a specific relationship. For example, the formula $d = rt$ represents the relationship between distance, rate, and time. If Bob travels at a rate of 35 miles per hour on his road trip from Westminster to Seneca, the formula $d = 35t$ can be used to represent his distance traveled in a specific length of time. Formulas can also be used to show different roles of the variables, transformed without any given numbers. Solving for r, the formula becomes $\frac{d}{t} = r$. The t is moved over by division so that **rate** is a function of distance and time.

Features of Graphs and Tables for Linear and Nonlinear Relationships

As mentioned, linear relationships describe the way two quantities change with respect to each other. The relationship is defined as linear because a line is produced if all the sets of corresponding values are graphed on a coordinate grid. When expressing the linear relationship as an equation, the equation is often written in the form $y = mx + b$ (**slope-intercept form**) where *m* and *b* are numerical values and *x* and *y* are variables (for example, $y = 5x + 10$). The slope is the coefficient of x, and the y-intercept is the constant value. The slope of the line containing the same two points is:

$$m = \frac{y_2 - y_1}{x_2 - x_1}$$

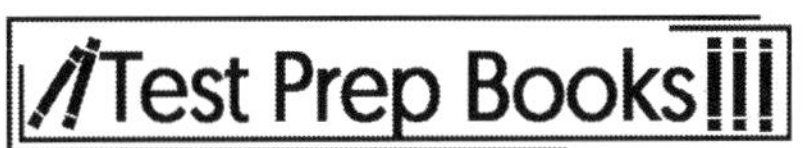

and is also equal to rise/run. Given a linear equation and the value of either variable (*x* or *y*), the value of the other variable can be determined.

With polynomial functions such as quadratics, the x-intercepts represent zeros of the function. Finding the **zeros of polynomial functions** is the same process as finding the solutions of polynomial equations. These are the points at which the graph of the function crosses the x-axis. In the following quadratic equation, factoring the binomial leads to finding the zeros of the function:

$$x^2 - 5x + 6 = y$$

This equations factors into

$$(x - 3)(x - 2) = y$$

where 2 and 3 are found to be the zeros of the function when y is set equal to zero. The zeros of any function are the x-values where the graph of the function on the coordinate plane crosses the x-axis, which is the same as an x-intercept.

Selecting an Equation that Best Represents a Graph

Three common functions used to model different relationships between quantities are linear, quadratic, and exponential functions. **Linear functions** are the simplest of the three, and the independent variable x has an exponent of 1. Written in the most common form:

$$y = mx + b$$

the coefficient of *x* indicates how fast the function grows at a constant rate, and the *b*-value denotes the starting point. A **quadratic function** has an exponent of 2 on the independent variable x. Standard form for this type of function is:

$$y = ax^2 + bx + c$$

and the graph is a parabola. These type functions grow at a changing rate. An **exponential function** has an independent variable in the exponent $y = ab^x$. The graph of these types of functions is described as **growth** or **decay**, based on whether the **base**, b, is greater than or less than 1. These functions are different from quadratic functions because the base stays constant. A common base is base *e*.

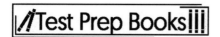

The following three functions model a linear, quadratic, and exponential function respectively: $y = 2x$, $y = x^2$, and $y = 2^x$. Their graphs are shown below. The first graph, modeling the linear function, shows that the growth is constant over each interval. With a horizontal change of 1, the vertical change is 2. It models constant positive growth. The second graph shows the quadratic function, which is a curve that is symmetric across the y-axis. The growth is not constant, but the change is mirrored over the axis. The last graph models the exponential function, where the horizontal change of 1 yields a vertical change that increases more and more with each iteration of horizontal change. The exponential graph gets very close to the x-axis, but never touches it, meaning there is an asymptote there. The y-value can never be zero because the base of 2 can never be raised to an input value that yields an output of zero.

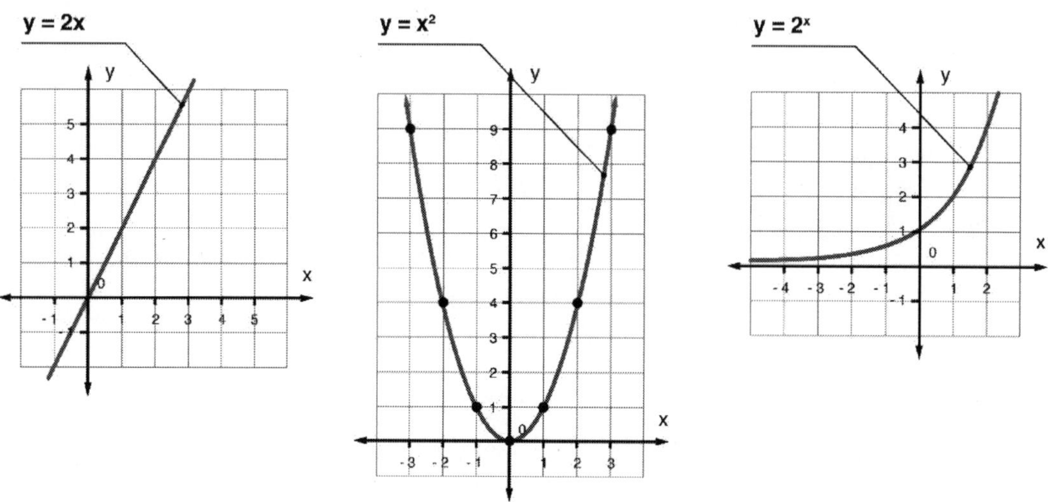

Determining the Graphical Properties and Sketch a Graph Given an Equation

Graphing a Linear Function

The process for graphing a line depends on the form in which its equation is written: slope-intercept form or standard form.

When an equation is written in slope-intercept form, $y = mx + b$, *m* represents the slope of the line and *b* represents the *y*-intercept. The *y*-intercept is the value of *y* when $x = 0$ and the point at which the graph of the line crosses the *y*-axis. The slope is the rate of change between the variables, expressed as a fraction. The fraction expresses the change in *y* compared to the change in *x*. If the slope is an integer, it should be written as a fraction with a denominator of 1. For example, 5 would be written as $\frac{5}{1}$.

To graph a line given an equation in slope-intercept form, the *y*-intercept should first be plotted. For example, to graph:

$$y = -\frac{2}{3}x + 7$$

the *y*-intercept of 7 would be plotted on the *y*-axis (vertical axis) at the point (0, 7). Next, the slope would be used to determine a second point for the line. Note that all that is necessary to graph a line is two points on that line. The slope will indicate how to get from one point on the line to another. The slope expresses vertical change (*y*) compared to horizontal change (*x*) and therefore is sometimes

referred to as $\frac{rise}{run}$. The numerator indicates the change in the *y* value (move up for positive integers and move down for negative integers), and the denominator indicates the change in the *x* value. For the previous example, using the slope of $-\frac{2}{3}$, from the first point at the *y*-intercept, the second point should be found by counting down 2 and to the right 3. This point would be located at (3, 5).

When an equation is written in standard form, $Ax + By = C$, it is easy to identify the *x*- and *y*-intercepts for the graph of the line. Just as the *y*-intercept is the point at which the line intercepts the *y*-axis, the *x*-intercept is the point at which the line intercepts the *x*-axis. At the *y*-intercept, $x = 0$; and at the *x*-intercept, $y = 0$. Given an equation in standard form, $x = 0$ should be used to find the *y*-intercept. Likewise, $y = 0$ should be used to find the *x*-intercept. For example, to graph $3x + 2y = 6$, 0 for *y* results in $3x + 2(0) = 6$. Solving for *y* yields $x = 2$; therefore, an ordered pair for the line is (2, 0). Substituting 0 for *x* results in $3(0) + 2y = 6$. Solving for *y* yields $y = 3$; therefore, an ordered pair for the line is (0, 3). The two ordered pairs (the *x*- and *y*-intercepts) can be plotted, and a straight line through them can be constructed.

T - chart

x	y
0	3
2	0

Intercepts

x - intercept : (2,0)

y - intercept : (0,3)

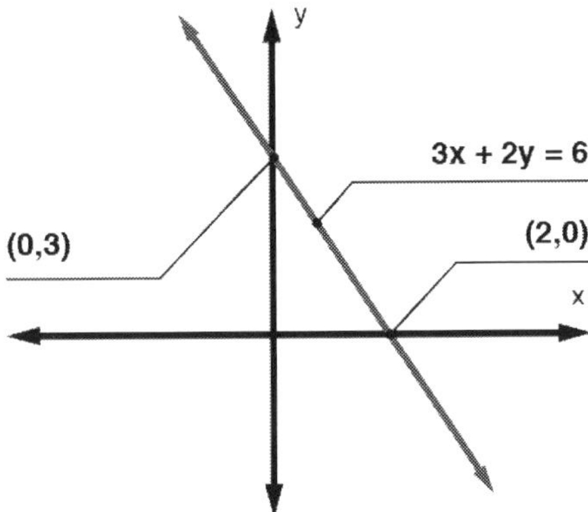

Graphing a Quadratic Function

The standard form of a quadratic function is:

$$y = ax^2 + bx + c$$

The graph of a quadratic function is a u-shaped (or upside-down u) curve, called a parabola, which is symmetric about a vertical line (axis of symmetry). To graph a parabola, its vertex (high or low point for the curve) and at least two points on each side of the axis of symmetry need to be determined.

Given a quadratic function in standard form, the axis of symmetry for its graph is the line $x = -\frac{b}{2a}$. The vertex for the parabola has an x-coordinate of $-\frac{b}{2a}$. To find the y-coordinate for the vertex, the calculated x-coordinate needs to be substituted. To complete the graph, two different x-values need to be selected and substituted into the quadratic function to obtain the corresponding y-values. This will give two points on the parabola. These two points and the axis of symmetry are used to determine the two points corresponding to these. The corresponding points are the same distance from the axis of symmetry (on the other side) and contain the same y-coordinate. Plotting the vertex and four other points on the parabola allows for constructing the curve.

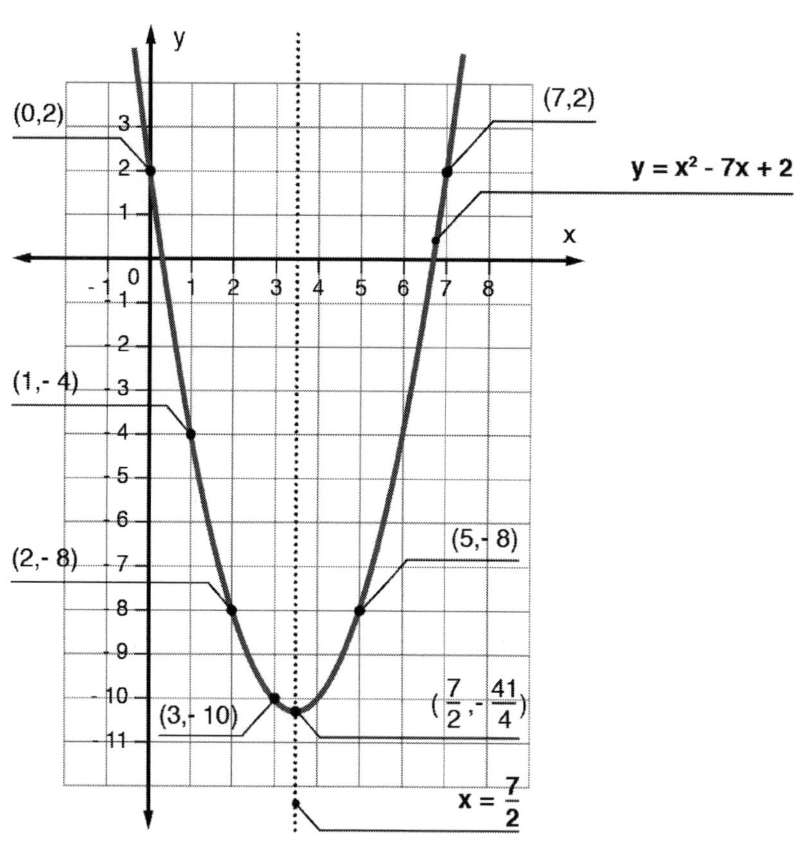

Graphing an Exponential Function

Exponential functions have a general form of:

$$y = a \times b^x$$

The graph of an exponential function is a curve that slopes upward or downward from left to right. The graph approaches a line, called an asymptote, as *x* or *y* increases or decreases. To graph the curve for an exponential function, *x*-values are selected and then substituted into the function to obtain the corresponding *y*-values. A general rule of thumb is to select three negative values, zero, and three positive values. Plotting the seven points on the graph for an exponential function should allow for constructing a smooth curve through them.

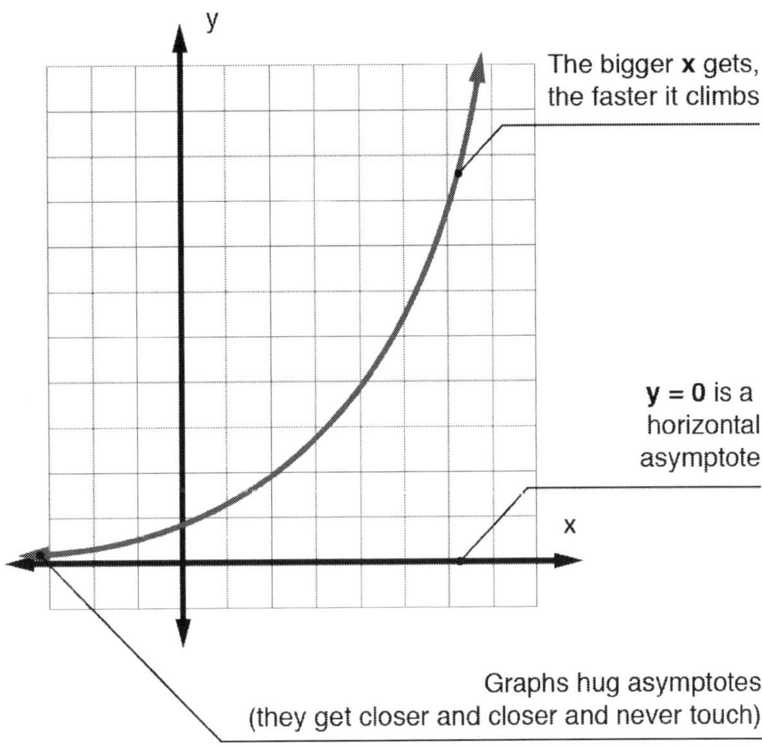
y
The bigger **x** gets, the faster it climbs
y = 0 is a horizontal asymptote
x
Graphs hug asymptotes (they get closer and closer and never touch)

Equation of a Line from the Slope and a Point on a Line

The point-slope form of a line:

$$y - y_1 = m(x - x_1)$$

is used to write an equation when given an ordered pair (point on the equation's graph) for the function and its rate of change (slope of the line). The values for the slope, *m*, and the point (x_1, y_1) are substituted into the point-slope form to obtain the equation of the line. A line with a slope of 3 and an ordered pair (4, –2) would have an equation:

$$y - (-2) = 3(x - 4)$$

If a question specifies that the equation be written in slope-intercept form, the equation should be manipulated to isolate *y*:

Solve: $y - (-2) = 3(x - 4)$

Distribute: $y + 2 = 3x - 12$

Subtract 2 from both sides: $y = 3x - 14$

Equation of a Line from Two Points

Given two ordered pairs for a function, (x_1, y_1) and (x_2, y_2), it is possible to determine the rate of change between the variables (slope of the line). To calculate the slope of the line, m, the values for the ordered pairs should be substituted into the formula:

$$m = \frac{y_2 - y_1}{x_2 - x_1}$$

The expression is substituted to obtain a whole number or fraction for the slope. Once the slope is calculated, the slope and either of the ordered pairs should be substituted into the point-slope form to obtain the equation of the line.

Using Slope of a Line

Two lines are parallel if they have the same slope and a different intercept. Two lines are **perpendicular** if the product of their slope equals –1. Parallel lines never intersect unless they are the same line, and perpendicular lines intersect at a right angle. If two lines aren't parallel, they must intersect at one point. If lines do cross, they're labeled as **intersecting lines** because they "intersect" at one point. If they intersect at more than one point, they're the same line. Determining equations of lines based on properties of parallel and perpendicular lines appears in word problems. To find an equation of a line, both the slope and a point the line goes through are necessary. Therefore, if an equation of a line is needed that's parallel to a given line and runs through a specified point, the slope of the given line and the point are plugged into the point-slope form of an equation of a line. Secondly, if an equation of a line is needed that's perpendicular to a given line running through a specified point, the negative reciprocal of the slope of the given line and the point are plugged into the **point-slope form**. Also, if the point of intersection of two lines is known, that point will be used to solve the set of equations. Therefore, to solve a system of equations, the point of intersection must be found. If a set of two equations with two unknown variables has no solution, the lines are parallel.

The **Parallel Postulate** states that if two parallel lines are cut by a transversal, then the corresponding angles are equal. Here is a picture that highlights this postulate:

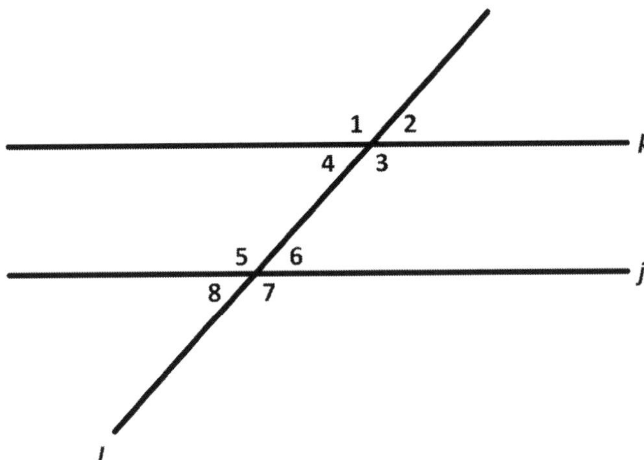

Because lines *k* and *i* are parallel, when cut by transversal *l,* angles 1 and 5 are equal, angles 2 and 6 are equal, angles 4 and 8 are equal, and angles 3 and 7 are equal. Note that angles 1 and 2, 3 and 4, 5 and 6, and 7 and 8 add up to 180 degrees.

This statement is equivalent to the **Alternate Interior Angle Theorem**, which states that when two parallel lines are cut by a transversal, the resultant interior angles are congruent. In the picture above, angles 3 and 5 are congruent, and angles 4 and 6 are congruent.

The Parallel Postulate or the Alternate Interior Angle Theorem can be used to find the missing angles in the following picture:

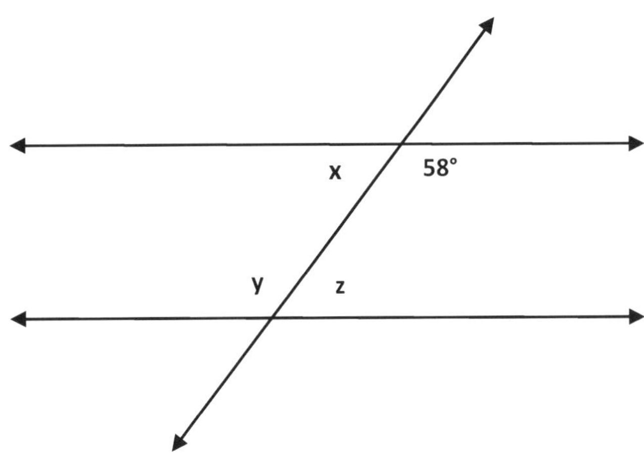

Assuming that the lines are parallel, angle x is found to be 122 degrees. Angle x and the 58-degree angle add up to 180 degrees. The Alternate Interior Angle Theorem states that angle y is equal to 58 degrees. Also, angles y and z add up to 180 degrees, so angle z is 122 degrees. Note that angles x and z are also alternate interior angles, so their equivalence can be used to find angle z as well.

An equivalent statement to the Parallel Postulate is that the sum of all angles in a triangle is 180 degrees. Therefore, given any triangle, if two angles are known, the third can be found accordingly.

Functions Shown in Different Ways

First, it's important to understand the definition of a **relation**. Given two variables, *x* and *y*, which stand for unknown numbers, a **relation** between *x* and *y* is an object that splits all of the pairs (*x*, *y*) into those for which the relation is true and those for which it is false. For example, consider the relation of $x^2 = y^2$. This relationship is true for the pair (1, 1) and for the pair (–2, 2), but false for (2, 3). Another example of a relation is $x \leq y$. This is true whenever *x* is less than or equal to *y*.

A **function** is a special kind of relation where, for each value of *x*, there is only a single value of *y* that satisfies the relation. So, $x^2 = y^2$ is *not* a function because in this case, if *x* is 1, *y* can be either 1 or –1: the pair (1, 1) and (1, –1) both satisfy the relation. More generally, for this relation, any pair of the form $(a, \pm a)$ will satisfy it. On the other hand, consider the following relation:

$$y = x^2 + 1$$

This is a function because for each value of *x*, there is a unique value of *y* that satisfies the relation. Notice, however, there are multiple values of *x* that give us the same value of *y*. This is perfectly acceptable for a function. Therefore, *y* is a function of *x*.

To determine if a relation is a function, check to see if every *x* value has a unique corresponding *y* value.

A function can be viewed as an object that has *x* as its input and outputs a unique *y*-value. It is sometimes convenient to express this using **function notation**, where the function itself is given a name, often *f*. To emphasize that *f* takes *x* as its input, the function is written as $f(x)$. In the above example, the equation could be rewritten as:

$$f(x) = x^2 + 1$$

To write the value that a function yields for some specific value of *x*, that value is put in place of *x* in the function notation. For example, $f(3)$ means the value that the function outputs when the input value is 3. If:

$$f(x) = x^2 + 1$$

then:

$$f(3) = 3^2 + 1 = 10$$

Another example of a function would be:

$$f(x) = 4x + 4$$

read "*f* of *x* is equal to four times *x* plus four." In this example, the input would be *x* and the output would be f(*x*). Ordered pairs would be represented as (*x*, f(*x*)). To find the output for an input value of 3, 3 would be substituted for *x* into the function as follows:

$$f(3) = 4(3) + 4$$

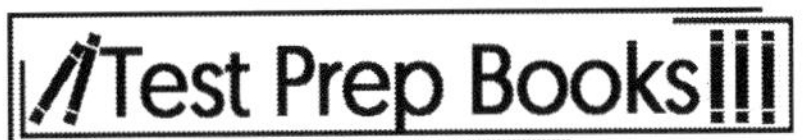

resulting in $f(3) = 16$. Therefore, the ordered pair:

$$(3, f(3)) = (3, 16)$$

Note f(*x*) is a function of *x* denoted by *f*. Functions of *x* could be named *g*(*x*), read "*g* of *x*"; *p*(*x*), read "*p* of *x*"; etc.

As an example, the following function is in function notation:

$$f(x) = 3x - 4$$

The $f(x)$ represents the output value for an input of x. If $x = 2$, the equation becomes:

$$f(2) = 3(2) - 4 = 6 - 4 = 2$$

The input of 2 yields an output of 2, forming the ordered pair $(2, 2)$. The following set of ordered pairs corresponds to the given function: $(2, 2), (0, -4), (-2, -10)$. The set of all possible inputs of a function is its **domain**, and all possible outputs is called the **range**. By definition, each member of the domain is paired with only one member of the range.

Functions can also be defined recursively. In this form, they are not defined explicitly in terms of variables. Instead, they are defined using previously-evaluated function outputs, starting with either $f(0)$ or $f(1)$. An example of a recursively-defined function is:

$$f(1) = 2, f(n) = 2f(n - 1) + 2n, n > 1$$

The domain of this function is the set of all integers.

A function can also be viewed as a table of pairs (*x*, *y*), which lists the value for *y* for each possible value of *x*.

Functions in Tables and Graphs

The domain and range of a function can be found by observing a table. The table below shows the input values $x = -2$ to $x = 2$ for the function:

$$f(x) = x^2 - 3$$

The range, or output, for these inputs results in a minimum of -3. On each side of $x = 0$, the numbers increase, showing that the range is all real numbers greater than or equal to -3.

x (domain/input)	y (range/output)
–2	1
–1	–2
0	–3
–1	–2
2	1

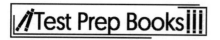

Determining the Domain and Range from a Given Graph of a Function

The domain and range of a function can also be found visually by its plot on the coordinate plane. In the function:

$$f(x) = x^2 - 3$$

for example, the domain is all real numbers because the parabola can stretch infinitely far left and right with no restrictions. This means that any input value from the real number system will yield an output in the real number system. For the range, the inequality $y \geq -3$ would be used to describe the possible output values because the parabola has a minimum at $y = -3$. This means there will not be any real output values less than -3 because -3 is the lowest value the function reaches on the y-axis.

Determining the Domain and Range of a Given Function

The set of all possible values for x in $f(x)$ is called the **domain** of the function, and the set of all possible outputs is called the **range** of the function. Note that usually the domain is assumed to be all real numbers, except those for which the expression for $f(x)$ is not defined, unless the problem specifies otherwise. An example of how a function might not be defined is in the case of:

$$f(x) = \frac{1}{x+1}$$

which is not defined when $x = -1$ (which would require dividing by zero). Therefore, in this case the domain would be all real numbers except $x = -1$.

Interpreting Domain and Range in Real-World Settings

A function can be built from the information given in a situation. For example, the relationship between the money paid for a gym membership and the number of months that someone has been a member can be described through a function. If the one-time membership fee is $40 and the monthly fee is $30, then the function can be written:

$$f(x) = 30x + 40$$

The x-value represents the number of months the person has been part of the gym, while the output is the total money paid for the membership. The table below shows this relationship. It is a representation of the function because the initial cost is $40 and the cost increases each month by $30.

x (months)	**f(x) (money paid to gym)**
0	40
1	70
2	100
3	130

In this situation, the domain of the function is real numbers greater than or equal to zero because it represents the number of months that a membership is held. We aren't told if the gym prorates memberships for partial months (if you join 10 days into a month, for example). If not, the domain would only be whole numbers plus zero, since there's a meaningful data point of $40, as a fee for joining. The range is real numbers greater than or equal to 40, because the range represents the total cost of the gym membership. Because there is a one-time fee of $40, the cost of carrying a membership will never be less than $40, so this is the minimum value.

When working through any word problem, the domain and range of the function should be considered in terms of the real-world context that the function models. For example, considering the above function for the cost of a gym membership, it would be nonsensical to include negative numbers in either the domain or range because there can't be negative months that someone holds a membership and similarly, the gym isn't going to pay a person for months prior to becoming a member. Therefore, while the function to model the situation (defined as $f(x) = 30x + 40$) theoretically could result in a true mathematical statement if negative values of are inputted, this would not make sense in the real-world context for which the function applies. Therefore, defining the domain as whole numbers and the range as all real numbers greater than or equal to 40 is important.

Evaluating Functions

To evaluate functions, plug in the given value everywhere the variable appears in the expression for the function. For example, find $f(-2)$ where:

$$f(x) = 2x^2 - \frac{4}{x}$$

To complete the problem, plug in –2 in the following way:

$$f(-2) = 2(-2)^2 - \frac{4}{-2}$$

$$2 \times 4 + 2$$

$$8 + 2 = 10$$

GED Mathematical Reasoning Practice Test #1

1. Which of the following is the result of simplifying the expression:

$$\frac{4a^{-1}b^3}{a^4b^{-2}} \times \frac{3a}{b}$$

a. $12a^3b^5$

b. $12\frac{b^4}{a^4}$

c. $\frac{12}{a^4}$

d. $7\frac{b^4}{a}$

2. What is the product of two irrational numbers?

a. Irrational

b. Rational

c. Contradictory

d. Irrational or rational

3. The graph shows the position of a car over a 10-second time interval. Which of the following is the correct interpretation of the graph for the interval 1 to 3 seconds?

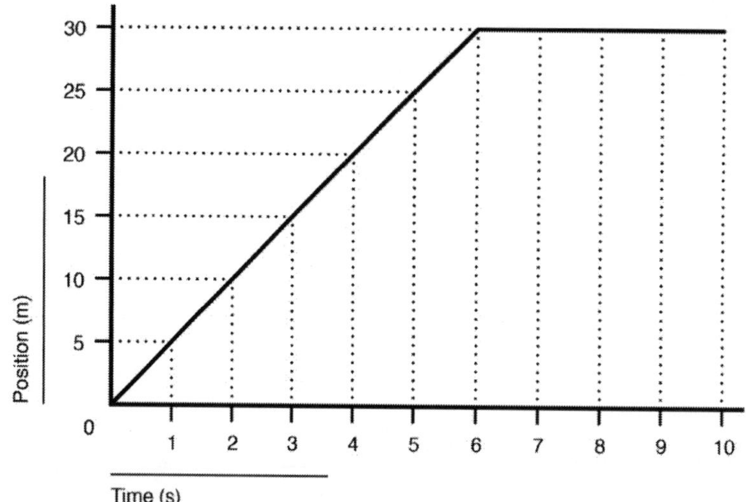

a. The car remains in the same position.

b. The car is traveling at a speed of 5 m/s.

c. The car is traveling up a hill.

d. The car is traveling at 5 mph.

4. How is the number -4 classified?
   a. Real, rational, integer, whole, natural
   b. Real, rational, integer, natural
   c. Real, rational, integer
   d. Real, irrational

5. In a statistical experiment, 29 college students are given an exam during week 11 of the semester, and 30 college students are given an exam during week 12 of the semester. Both groups are being tested to determine which exam week might result in a higher grade. What's the degree of freedom in this experiment?
   a. 29
   b. 30
   c. 59
   d. 28

6. What are the zeros of the function: $f(x) = x^3 + 4x^2 + 4x$?
   a. -2
   b. $0, -2$
   c. 2
   d. $0, 2$

7. If $g(x) = x^3 - 3x^2 - 2x + 6$ and $f(x) = 2$, then what is $g(f(x))$?
   a. -26
   b. 6
   c. $2x^3 - 6x^2 - 4x + 12$
   d. -2

8. $(2x - 4y)^2 =$
   a. $4x^2 - 16xy + 16y^2$
   b. $4x^2 - 8xy + 16y^2$
   c. $4x^2 - 16xy - 16y^2$
   d. $2x^2 - 8xy + 8y^2$

9. If $x \neq 0$, then $\frac{3}{x} + \frac{5u}{2x} - \frac{u}{4} =$
   a. $\frac{12+10u-ux}{4x}$
   b. $\frac{3+5u-ux}{x}$
   c. $\frac{12x+10u+ux}{4x}$
   d. $\frac{12+10u-u}{4x}$

10. What is the product of the following expression?

$$(4x-8)(5x^2+x+6)$$

a. $20x^3-36x^2+16x-48$
b. $6x^3-41x^2+12x+15$
c. $20x^4+11x^2-37x-12$
d. $2x^3-11x^2-32x+20$

11. How could the following equation be factored to find the zeros?

$$y=x^3-3x^2-4x$$

a. $0=x^2(x-4), x=0,4$
b. $0=3x(x+1)(x+4), x=0,-1,-4$
c. $0=x(x+1)(x+6), x=0,-1,-6$
d. $0=x(x+1)(x-4), x=0,-1,4$

12. What is the simplified quotient of $\frac{5x^3}{3x^2y} \div \frac{25}{3y^9}$?

a. $\frac{125x}{9y^{10}}$
b. $\frac{x}{5y^8}$
c. $\frac{5}{xy^8}$
d. $\frac{xy^8}{5}$

13. What is the solution for the following equation?

$$\frac{x^2+x-30}{x-5}=11$$

a. $x=-6$
b. There is no solution.
c. $x=16$
d. $x=5$

14. Mom's car drove 72 miles in 90 minutes. How fast did she drive in feet per second?
a. 0.8 feet per second
b. 48.9 feet per second
c. 0.009 feet per second
d. 70.4 feet per second

15. How do you solve $V = lwh$ for h?
   a. $lwV = h$
   b. $h = \frac{V}{lw}$
   c. $h = \frac{Vl}{w}$
   d. $h = \frac{Vw}{l}$

16. What is the domain for the function $y = \sqrt{x}$?
   a. All real numbers
   b. $x \geq 0$
   c. $x > 0$
   d. $y \geq 0$

17. If Sarah reads at an average rate of 21 pages in four nights, how long will it take her to read 140 pages?
   a. 6 nights
   b. 26 nights
   c. 8 nights
   d. 27 nights

18. The phone bill is calculated each month using the equation $c = 50g + 75$. The cost of the phone bill per month is represented by c, and g represents the gigabytes of data used that month. What is the value and interpretation of the slope of this equation?
   a. 75 dollars per day
   b. 75 gigabytes per day
   c. 50 dollars per day
   d. 50 dollars per gigabyte

19. What is the function that forms an equivalent graph to $y = \cos(x)$?
   a. $y = \tan(x)$
   b. $y = \csc(x)$
   c. $y = \sin(x + \frac{\pi}{2})$
   d. $y = \sin(x - \frac{\pi}{2})$

20. What is the solution for the equation $\tan(x) + 1 = 0$, where $0 \leq x < 2\pi$?
   a. $x = \frac{3\pi}{4}, \frac{5\pi}{4}$
   b. $x = \frac{3\pi}{4}, \frac{\pi}{4}$
   c. $x = \frac{5\pi}{4}, \frac{7\pi}{4}$
   d. $x = \frac{3\pi}{4}, \frac{7\pi}{4}$

21. What is the inverse of the function $f(x) = 3x - 5$?
   a. $f^{-1}(x) = \frac{x}{3} + 5$
   b. $f^{-1}(x) = \frac{5x}{3}$
   c. $f^{-1}(x) = 3x + 5$
   d. $f^{-1}(x) = \frac{x+5}{3}$

22. What are the zeros of $f(x) = x^2 + 4$?
   a. $x = -4$
   b. $x = \pm 2i$
   c. $x = \pm 2$
   d. $x = \pm 4i$

23. Twenty is 40 percent of what number?
   a. 5,000
   b. 8
   c. 200
   d. 50

24. What is the simplified form of the expression $1.2 \times 10^{12} \div 3.0 \times 10^8$?
   a. 0.4×10^4
   b. 4.0×10^4
   c. 4.0×10^3
   d. 3.6×10^{20}

25. You measure the width of your door to be 36 inches. The true width of the door is 35.75 inches. What is the relative error in your measurement?
   a. 0.7%
   b. 0.007%
   c. 0.99%
   d. 0.1%

26. What is the y-intercept for $y = x^2 + 3x - 4$?
   a. $y = 1$
   b. $y = -4$
   c. $y = 3$
   d. $y = 4$

27. On Monday, Robert mopped the floor in 4 hours. On Tuesday, he did it in 3 hours. If on Monday, his average rate of mopping was p sq. ft. per hour, what was his average rate on Tuesday?
   a. $\frac{4}{3}p$ sq. ft. per hour
   b. $\frac{3}{4}p$ sq. ft. per hour
   c. $\frac{5}{4}p$ sq. ft. per hour
   d. $p + 1$ sq. ft. per hour

28. Which equation is not a function?
   a. $y = |x|$
   b. $y = \frac{1}{x}$
   c. $x = 3$
   d. $y = 4$

29. How could the following function be rewritten to identify the zeros?

$$y = 3x^3 + 3x^2 - 18x$$

a. $y = 3x(x + 3)(x - 2)$
b. $y = x(x - 2)(x + 3)$
c. $y = 3x(x - 3)(x + 2)$
d. $y = (x + 3)(x - 2)$

30. What is the slope of the line tangent to the graph of $y = x^3 - 4$ at the point where $x = 2$?
a. $3x^2$
b. 4
c. -4
d. 12

31. Given the following triangle, what's the length of the missing side? Round the answer to the nearest tenth.

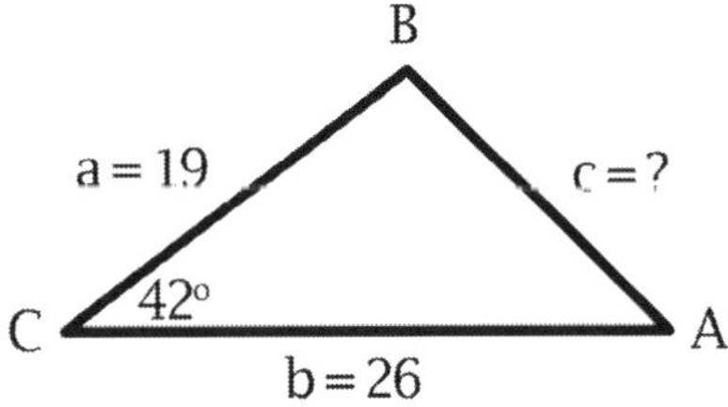

a. 17.0
b. 17.4
c. 18.0
d. 18.4

32. How many possible two-number pairs are there for the numbers 1, 2, 3, 4, and 5 if each number can only be used once and order DOES matter?
a. 120
b. 60
c. 20
d. 10

33. What are the first four terms of the series $\left\{\frac{(-1)^{n+1}}{n^2+5}\right\}_{n=0}^{\infty}$?
a. $\frac{1}{6}, \frac{1}{9}, \frac{1}{14}, \frac{1}{19}$
b. $\frac{1}{6}, \frac{-1}{9}, \frac{1}{14}, \frac{-1}{19}$
c. $\frac{-1}{5}, \frac{1}{6}, \frac{-1}{9}, \frac{1}{14}$
d. $\frac{1}{5}, \frac{1}{6}, \frac{1}{9}, \frac{1}{14}$

34. A particle moves along the x-axis, so that at any time $t \geq 0$, its velocity is given by $v(t) = \frac{6}{t+3}$. What is the acceleration of the particle at time $t = 5$?

a. $-\frac{2}{3}$
b. $-\frac{3}{32}$
c. $\frac{3}{4}$
d. $\frac{2}{3}$

35. If the volume of a sphere is 288π cubic meters, what are the radius and surface area of the same sphere?

a. Radius: 6 meters, Surface Area: 144π square meters
b. Radius: 36 meters, Surface Area: 144π square meters
c. Radius: 6 meters, Surface Area: 12π square meters
d. Radius: 36 meters, Surface Area: 12π square meters

36. The triangle shown below is a right triangle. What's the value of x?

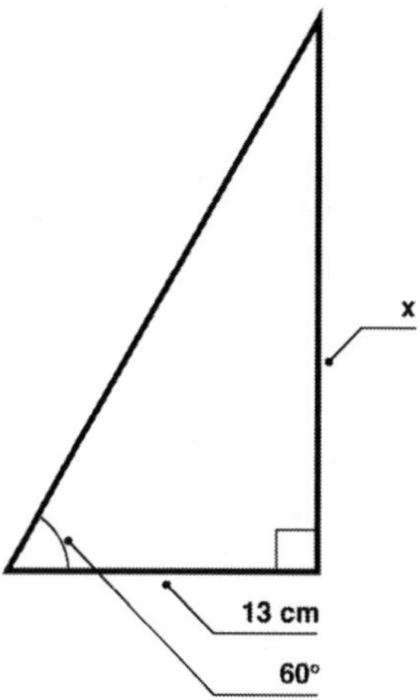

a. $x = 1.73$
b. $x = 0.57$
c. $x = 13$
d. $x = 22.52$

37. What's the midpoint of a line segment with endpoints $(-1, 2)$ and $(3, -6)$?

a. $(1, 2)$
b. $(1, 0)$
c. $(-1, 2)$
d. $(1, -2)$

38. What is the type of function that is modeled by the values in the following table?

X	f(x)
1	2
2	4
3	8
4	16
5	32

a. Linear
b. Exponential
c. Quadratic
d. Cubic

39. What is the simplified form of the following expression $tan\theta\ cos\theta$?
a. $sin\theta$
b. 1
c. $csc\theta$
d. $tan\theta$

40. A sample data set contains the following values: 1, 3, 5, 7. What's the standard deviation of the set?
a. 2.58
b. 4
c. 6.23
d. 1.1

41. A pair of dice is thrown, and the sum of the two scores is calculated. What's the expected value of the roll?
a. 5
b. 6
c. 7
d. 8

42. A ball is drawn at random from a ball pit containing 8 red balls, 7 yellow balls, 6 green balls, and 5 purple balls. What's the probability that the ball drawn is yellow?
a. $\frac{1}{26}$
b. $\frac{19}{26}$
c. $\frac{7}{26}$
d. 1

43. Two cards are drawn from a shuffled deck of 52 cards. What's the probability that both cards are kings if the first card isn't replaced after it's drawn?

a. $\frac{1}{169}$

b. $\frac{1}{221}$

c. $\frac{1}{13}$

d. $\frac{4}{13}$

44. What's the probability of rolling a 6 exactly once in two rolls of a die?

a. $\frac{1}{3}$

b. $\frac{1}{36}$

c. $\frac{1}{6}$

d. $\frac{11}{36}$

45. Given the set $A = \{1, 2, 3, 4, 5, 6, 7, 8, 9, 10\}$ and $B = \{1, 2, 3, 4, 5\}$, find $A - (A \cap B)$.

a. $\{6, 7, 8, 9, 10\}$
b. $\{1, 2, 3, 4, 5\}$
c. $\{1, 2, 3, 4, 5, 6, 7, 8, 9, 10\}$
d. $\emptyset$

46. Let p equal "Alex is an engineering major," q equal "Alex is not an English major," r equal "Alex's sister is a history major," s equal "Alex's sister has been to Germany," and t equal "Alex's sister has been to Austria." Which of the following answers represents the statement, "Alex is an engineering and English major, but his sister is a history major who hasn't been to either Germany or Austria"?

a. $p \wedge \sim q \wedge (r \vee (\sim s \vee \sim t))$
b. $p \wedge q \wedge r \vee (\sim s \wedge \sim t)$
c. $p \wedge \sim q \wedge r \wedge (\sim s \vee \sim t)$
d. $p \wedge q \wedge (r \vee (\sim s \wedge \sim t))$

Answer Explanations for Practice Test #1

1. B: To simplify the given equation, the first step is to make all exponents positive by moving them to the opposite place in the fraction. This expression becomes:

$$\frac{4b^3b^2}{a^1a^4} \times \frac{3a}{b}$$

Then the rules for exponents can be used to simplify. Multiplying the same bases means the exponents can be added. Dividing the same bases means the exponents are subtracted. Thus, after multiplying the exponents in the first fraction, the equation becomes:

$$\frac{4b^5}{a^5} \times \frac{3a}{b}$$

Therefore, we can first multiply to get:

$$\frac{12ab^5}{a^5b}$$

Then, simplifying yields:

$$12\frac{b^4}{a^4}$$

2. D: The product of two irrational numbers can be rational or irrational. Sometimes the irrational parts of the two numbers cancel each other out, leaving a rational number. For example, $\sqrt{2} \times \sqrt{2} = 2$ because the roots cancel each other out. Technically, the product of two irrational numbers can be complex because complex numbers can have either the real or imaginary part (in this case, the imaginary part) equal zero and still be considered a complex number. However, Choice *D* is incorrect because the product of two irrational numbers is not an imaginary number so saying the product is complex *and* imaginary is incorrect.

3. B: The car is traveling at a speed of five meters per second. On the interval from one to three seconds, the position changes by ten meters. By making this change in position over time into a rate, the speed becomes ten meters in two seconds, or five meters in one second.

4. C: The number negative four is classified as a real number because it exists and is not imaginary. It is rational because it does not have a decimal that never ends. It is an integer because it does not have a fractional component. The next classification would be whole numbers, for which negative four does not qualify because it is negative.

5. D: The degree of freedom for two samples is calculated as $df = \frac{(n_1-1)+(n_2-1)}{2}$ rounded to the lowest whole number. For this example, $df = \frac{(29-1)+(30-1)}{2} = \frac{28+29}{2} = 28.5$ which, rounded to the lowest whole number, is 28.

6. B: There are two zeros for the function $x = 0, -2$. The zeros can be found several ways, but this particular equation can be factored into:

$$f(x) = x(x^2 + 4x + 4) = x(x + 2)(x + 2)$$

By setting each factor equal to zero and solving for x, there are two solutions. On a graph, these zeros can be seen where the line crosses the x-axis.

7. D: This problem involves a composition function, where one function is plugged into the other function. In this case, the $f(x)$ function is plugged into the $g(x)$ function for each x-value. The composition equation becomes:

$$g(f(x)) = (2)^3 - 3(2)^2 - 2(2) + 6$$

Simplifying the equation gives the answer:

$$g(f(x)) = 8 - 3(4) - 2(2) + 6$$

$$g(f(x)) = 8 - 12 - 4 + 6$$

$$g(f(x)) = -2$$

8. A: To expand a squared binomial, it's necessary to use the First, Outer, Inner, Last (FOIL) Method.

$$(2x - 4y)^2$$

$$(2x)(2x) + (2x)(-4y) + (-4y)(2x) + (-4y)(-4y)$$

$$4x^2 - 8xy - 8xy + 16y^2$$

$$4x^2 - 16xy + 16y^2$$

9. A: The common denominator here will be $4x$. Rewrite these fractions as

$$\frac{3}{x} + \frac{5u}{2x} - \frac{u}{4} = \frac{12}{4x} + \frac{10u}{4x} - \frac{ux}{4x}$$

$$\frac{12x + 10u - ux}{4x}$$

10. A: Finding the product means distributing one polynomial onto the other. Each term in the first must be multiplied by each term in the second. Then, like terms can be collected. Multiplying the factors yields the expression:

$$20x^3 + 4x^2 + 24x - 40x^2 - 8x - 48$$

Collecting like terms means adding the x^2 terms and adding the x-terms. The final answer after simplifying the expression is:

$$20x^3 - 36x^2 + 16x - 48$$

11. D: Finding the zeros for a function by factoring is done by setting the equation equal to zero, then completely factoring. Since there is a common x for each term in the provided equation, that should be factored out first to get $x(x^2 - 3x - 4)$. Then the quadratic that is left can be factored into two binomials, which are $(x + 1)(x - 4)$. This gives the factored equation $0 = x(x + 1)(x - 4)$.

12. D: Dividing rational expressions follows the same rule as dividing fractions. The division is changed to multiplication by the reciprocal of the second fraction. This turns the expression into:

$$\frac{5x^3}{3x^2y} \times \frac{3y^9}{25}$$

This can be simplified by finding common factors in the numerators and denominators of the two fractions.

$$\frac{x^3}{x^2y} \times \frac{y^9}{5}$$

Multiplying across creates:

$$\frac{x^3y^9}{5x^2y}$$

Simplifying leads to the final expression of:

$$\frac{xy^8}{5}$$

13. B: The equation can be solved by factoring the numerator into $(x + 6)(x - 5)$. Since that same factor exists on top and bottom, that factor $(x - 5)$ cancels. This leaves the equation $x + 6 = 11$. Solving the equation gives the answer $x = 5$. When this value is plugged into the equation, it yields a zero in the denominator of the fraction. Since this is undefined, there is no solution.

14. D: This problem can be solved by using unit conversion. The initial units are miles per minute. The final units need to be feet per second. Converting miles to feet uses the equivalence statement $1 \text{ mi} = 5{,}280 \text{ ft}$. Converting minutes to seconds uses the equivalence statement $1 \text{ min} = 60 \text{ s}$. Setting up the ratios to convert the units is shown in the following equation:

$$\frac{72 \text{ mi}}{90 \text{ min}} \times \frac{1 \text{ min}}{60 \text{ s}} \times \frac{5{,}280 \text{ ft}}{1 \text{ mi}} = 70.4\frac{\text{ft}}{\text{s}}$$

The initial units cancel out, and the new units are left.

15. B: The formula can be manipulated by dividing both the length, l, and the width, w, on both sides. The length and width will cancel on the right, leaving height, h, by itself.

16. B: The domain is all possible input values, or x-values. For this equation, the domain is every number greater than or equal to zero. There are no negative numbers in the domain because taking the square root of a negative number results in an imaginary number.

17. D: This problem can be solved by setting up a proportion involving the given information and the unknown value. The proportion is:

$$\frac{21 \text{ pages}}{4 \text{ nights}} = \frac{140 \text{ pages}}{x \text{ nights}}$$

Solving the proportion by cross-multiplying, the equation becomes $21x = 4 \times 140$, where $x = 26.67$. Since it is not an exact number of nights, the answer is rounded up to 27 nights. Twenty-six nights would not give Sarah enough time.

18. D: The slope from this equation is 50, and it is interpreted as the cost per gigabyte used. Since the g-value represents number of gigabytes and the equation is set equal to the cost in dollars, the slope relates these two values. For every gigabyte used on the phone, the bill goes up 50 dollars.

19. C: Graphing the function $y = \cos(x)$ shows that the curve starts at $(0, 1)$, has an amplitude of 2, and a period of 2π. This same curve can be constructed using the sine graph, by shifting the graph to the left $\frac{\pi}{2}$ units. This equation is in the form $y = \sin(x + \frac{\pi}{2})$.

20. D: Using SOHCAHTOA, tangent is $\frac{y}{x}$ for the special triangles. Since the value of $\tan(x)$ needs to be -1, the angle for the tangent must be some form of 45 degrees or $\frac{\pi}{4}$. The value is negative in the second and fourth quadrant, so the answer is $\frac{3\pi}{4}$ and $\frac{7\pi}{4}$.

21. D: This inverse of a function is found by switching the x and y in the equation and solving for y. The given equation is $y = 3x - 5$, so solving for y is done by adding 5 to both sides, then dividing both sides by 3. This yields:

$$f^{-1}(x) = \frac{x + 5}{3}$$

This answer can be checked on the graph by verifying the lines are reflected over $y = x$.

22. B: The zeros of this function can be found by setting $f(x)$ equal to 0 and solving for x.

$$0 = x^2 + 4$$

$$-4 = x^2$$

$$\sqrt{-4} = x$$

Taking the square root of a negative number results in an imaginary number, so the solution is:

$$x = \pm 2i$$

23. D: Setting up a proportion is the easiest way to represent this situation. The proportion becomes $\frac{20}{x} = \frac{40}{100}$, where cross-multiplication can be used to solve for x. The answer can also be found by observing the two fractions as equivalent, knowing that twenty is half of forty, and fifty is half of one-hundred.

24. C: Scientific notation division can be solved by grouping the first terms together and grouping the tens together. The first terms can be divided, and the tens terms can be simplified using the rules for exponents. The initial expression becomes 0.4×10^4. This is not a positive whole number less than 10. Shifting the decimal and subtracting one from the exponent, the answer becomes 4.0×10^3.

25. A: The relative error can be found by finding the absolute error and making it a percent of the true value. The absolute error is $36 - 35.75 = 0.25$. This error is then divided by 35.75—the true value—to find 0.7%.

26. B: The y-intercept of an equation is found where the x-value is zero. Plugging zero into the equation for x, the first two terms cancel out, leaving -4.

27. D: If s is the size of the floor in square feet and r is the rate on Tuesday, then, based on the information given, $p = \frac{s}{4}$ and $r = \frac{s}{3}$. Solve the Monday rate for s, $s = 4p$, and then substitute that in the expression for Tuesday.

28. C: The equation $x = 3$ is not a function because it does not pass the vertical line test. This test is made from the definition of a function, where each x-value must be mapped to one and only one y-value. This equation is a vertical line, so the x-value of 3 is mapped with an infinite number of y-values.

29. A: The function can be factored to identify the zeros. First, the term $3x$ is factored out to the front because each term contains $3x$. Then, the quadratic is factored into $(x + 3)(x - 2)$.

30. D: Finding the slope of the line tangent to the given function involves taking the derivative twice. The first derivative gives the line tangent to the graph. The second derivative finds the slope of that line. The line tangent to the graph has an equation $y' = 3x^2$. The slope of this line at $x = 2$ is found by the second derivative, $y = 6x$, or $y = 6(2) = 12$.

31. B: Because this isn't a right triangle, SOHCAHTOA can't be used. However, the law of cosines can be used. Therefore,

$$c^2 = a^2 + b^2 - 2ab\cos C$$

$$c^2 = 19^2 + 26^2 - 2 \times 19 \times 26 \times \cos 42° = 302.773$$

Taking the square root and rounding to the nearest tenth results in $c = 17.4$.

32. C: Because order *does* matter, the total number of permutations needs to be computed.

$$P(5,2) = \frac{5!}{(5-2)!} = \frac{120}{6} = 20$$

20 represents the number of ways that two objects can be arranged from a set of five.

33. C: The numerator in the sequence $\left\{\frac{(-1)^{n+1}}{n^2+5}\right\}_{n=0}^{\infty}$ indicates that the sign of each term changes from term to term. The first term is negative because $n = 0$ and:

$$-1^{n+1} = -1^1 = -1$$

Therefore, the second term is positive. The third term is negative, etc. The denominator is evaluated like a function for plugging in various n values. For example, the denominator of the first term, when $n = 0$, is $0^2 + 5 = 5$.

34. B: The acceleration of the particle can be found by taking the derivative of the velocity equation. This equation is:

$$v'(t) = \frac{0 - 6(1)}{(t+3)^2} = \frac{-6}{(t+3)^2}$$

Finding the acceleration at time $t = 5$ can be found by plugging five in for the variable t in the derivative. The equation and answer are:

$$v'(5) = \frac{-6}{(5+3)^2} = \frac{-6}{64} = \frac{-3}{32}$$

35. A: Because the volume of the given sphere is 288π cubic meters, this gives:

$$\frac{4}{3}\pi r^3 = 288\pi$$

This equation is solved for r to obtain a radius of 6 meters. The formula for surface area is $4\pi r^2$, so:

$$SA = 4\pi 6^2 = 144\pi \text{ square meters}$$

36. D: SOHCAHTOA is used to find the missing side length. Because the angle and adjacent side are known, $\tan 60° = \frac{x}{13}$. Making sure to evaluate tangent with an argument in degrees, this equation gives:

$$x = 13 \tan 60 = 13 \times \sqrt{3} = 22.52$$

37. D: The midpoint formula should be used.

$$M = \left(\frac{x_1 + x_2}{2}, \frac{y_1 + y_2}{2}\right) = \left(\frac{-1+3}{2}, \frac{2+(-6)}{2}\right) = (1, -2)$$

38. B: The table shows values that are increasing exponentially. The differences between the inputs are the same, while the differences in the outputs are changing by a factor of 2. The values in the table can be modeled by the equation $f(x) = 2^x$.

39. A: Using the trigonometric identity $\tan(\theta) = \frac{\sin(\theta)}{\cos(\theta)}$, the expression becomes $\frac{\sin\theta}{\cos\theta}\cos\theta$. The factors that are the same on the top and bottom cancel out, leaving the simplified expression $\sin\theta$.

40. A: First, the sample mean must be calculated.

$$\bar{x} = \frac{1}{4}(1 + 3 + 5 + 7) = 4$$

The standard deviation of the data set is $s = \sqrt{\frac{\sum(x-\bar{x})^2}{n-1}}$, and $n = 4$ represents the number of data points.

Therefore,

$$s = \sqrt{\frac{1}{3}[(1-4)^2 + (3-4)^2 + (5-4)^2 + (7-4)^2]}$$

$$\sqrt{\frac{1}{3}(9+1+1+9)} = 2.58$$

41. C: To find the expected value, take the product of each individual sum and the probability of rolling the sum, then add together the products for each sum. There are 36 possible rolls.

The probability of rolling a 2 is $\frac{1}{36}$.

The probability of rolling a 3 is $\frac{2}{36}$.

The probability of rolling a 4 is $\frac{3}{36}$.

The probability of rolling a 5 is $\frac{4}{36}$.

The probability of rolling a 6 is $\frac{5}{36}$.

The probability of rolling a 7 is $\frac{6}{36}$.

The probability of rolling an 8 is $\frac{5}{36}$.

The probability of rolling a 9 is $\frac{4}{36}$.

The probability of rolling a 10 is $\frac{3}{36}$.

The probability of rolling an 11 is $\frac{2}{36}$.

Finally, the probability of rolling a 12 is $\frac{1}{36}$.

Each possible outcome is multiplied by the probability of it occurring. Like this:

$$2 \times \frac{1}{36} = a$$

$$3 \times \frac{2}{36} = b$$

$$4 \times \frac{3}{36} = c$$

And so forth.

Then, all of those results are added together:

$$a + b + c \ldots = expected\ value = \frac{252}{36}$$

In this case, it equals 7, which makes sense considering it is the value that has the highest probability of being rolled.

42. C: The sample space is made up of $8 + 7 + 6 + 5 = 26$ balls. The probability of pulling each individual ball is $\frac{1}{26}$. Since there are 7 yellow balls, the probability of pulling a yellow ball is $\frac{7}{26}$.

43. B: For the first card drawn, the probability of a king being pulled is $\frac{4}{52}$. Since this card isn't replaced, if a king is drawn first, the probability of a king being drawn second is $\frac{3}{51}$. The probability of a king being drawn in both the first and second draw is the product of the two probabilities:

$$\frac{4}{52} \times \frac{3}{51} = \frac{12}{2{,}652}$$

This fraction, when divided by $\frac{12}{12}$, equals $\frac{1}{221}$.

44. D: The addition rule is necessary to determine the probability because a 6 can be rolled on either roll of the die but not both. The rule used is:

$$P(A \text{ or } B) = P(A) + P(B) - P(A \text{ and } B)$$

The probability of a 6 being individually rolled is $\frac{1}{6}$ and the probability of a 6 being rolled twice is:

$$\frac{1}{6} \times \frac{1}{6} = \frac{1}{36}$$

Therefore, the probability that a 6 is rolled at least once is:

$$\frac{1}{6} + \frac{1}{6} - \frac{1}{36} = \frac{11}{36}$$

45. A: $(A \cap B)$ is equal to the intersection of the two sets A and B, which is $\{1, 2, 3, 4, 5\}$. $A - (A \cap B)$ is equal to the elements of A that are <u>not</u> included in the set $(A \cap B)$. Therefore:

$$A - (A \cap B) = \{6, 7, 8, 9, 10\}$$

46. C: "Alex is an engineering and English major, but his sister is a history major who hasn't been to either Germany or Austria" can be rewritten as "p and not q and r and not s or not t." Using logical symbols, this is written as $p \wedge {\sim}q \wedge r \wedge ({\sim}s \vee {\sim}t)$.

GED Mathematical Reasoning Practice Test #2

1. Which of the following is largest?
   a. 0.45
   b. 0.096
   c. 0.3
   d. 0.313

2. Which of the following is NOT a way to write 40 percent of N?
   a. $(0.4)N$
   b. $\frac{2}{5}N$
   c. $40N$
   d. $\frac{4N}{10}$

3. Which is closest to 17.8×9.9?
   a. 140
   b. 180
   c. 200
   d. 350

4. Five of six numbers have a sum of 25. The average of all six numbers is 6. What is the sixth number?
   a. 8
   b. 10
   c. 11
   d. 12

5. If $\frac{5}{2} \div \frac{1}{3} = n$, then n is between:
   a. 5 and 7
   b. 7 and 9
   c. 9 and 11
   d. 3 and 5

6. A closet is filled with red, blue, and green shirts. If $\frac{1}{3}$ of the shirts are green and $\frac{2}{5}$ are red, what fraction of the shirts are blue?
   a. $\frac{4}{15}$
   b. $\frac{1}{5}$
   c. $\frac{7}{15}$
   d. $\frac{1}{2}$

7. Shawna buys $2\frac{1}{2}$ gallons of paint. If she uses $\frac{1}{3}$ of it on the first day, how much does she have left?

a. $1\frac{5}{6}$ gallons

b. $1\frac{1}{2}$ gallons

c. $1\frac{2}{3}$ gallons

d. 2 gallons

8. On Monday, Robert mopped the floor in 4 hours. On Tuesday, he did it in 3 hours. If on Monday, his average rate of mopping was p sq. ft. per hour, what was his average rate on Tuesday?

a. $\frac{4}{3}p$ sq. ft. per hour

b. $\frac{3}{4}p$ sq. ft. per hour

c. $\frac{5}{4}p$ sq. ft. per hour

d. $p + 1$ sq. ft. per hour

9. The variable y is directly proportional to x. If $y = 3$ when $x = 5$, then what is y when $x = 20$?

a. 10
b. 12
c. 14
d. 16

10. There are $4x + 1$ treats in each party favor bag. If a total of $60x + 15$ treats are distributed, how many bags are given out?

a. 15
b. 16
c. 20
d. 22

11. A rectangle has a length that is 5 feet longer than three times its width. If the perimeter is 90 feet, what is the length in feet?

a. 10
b. 20
c. 25
d. 35

12. In an office, there are 50 workers. A total of 60% of the workers are women, and the chances of a woman wearing a skirt is 50%. If no men wear skirts, how many workers are wearing skirts?

a. 12
b. 15
c. 16
d. 20

13. What is the volume of a cube with the side equal to 3 inches?
   a. 6 in^3
   b. 27 in^3
   c. 9 in^3
   d. 3 in^3

14. What is the volume of a cube with the side equal to 5 centimeters?
   a. 10 cm^3
   b. 15 cm^3
   c. 50 cm^3
   d. 125 cm^3

15. What is the length of the hypotenuse of a right triangle with one leg equal to 3 centimeters and the other leg equal to 4 centimeters?
   a. 7 cm
   b. 5 cm
   c. 25 cm
   d. 12 cm

16. What is the length of the other leg of a right triangle with a hypotenuse of 10 inches and a leg of 8 inches?
   a. 6 in
   b. 18 in
   c. 80 in
   d. 13 in

17. What is the answer to $(2 + 2i)(2 - 2i)$?
   a. 8
   b. $8i$
   c. 4
   d. $4i$

18. What is the answer to $(3 + 3i)(3 - 3i)$?
   a. 18
   b. $18i$
   c. 9
   d. $9i$

19. What is the answer to $\frac{2+2i}{2-2i}$?
   a. 8
   b. $8i$
   c. $2i$
   d. i

20. What is the answer to $\frac{3+3i}{3-3i}$?

a. 18
b. $18i$
c. i
d. $9i$

21. According to building code regulations, the roof of a house has to be set at a minimum angle of 39° up to a maximum angle of 48° to ensure snow and rain will properly slide off it. What is the maximum incline in terms of radians?

a. $\frac{\pi}{4}$
b. $\frac{\pi}{15}$
c. $\frac{4\pi}{15}$
d. $\frac{3\pi}{4}$

22. Two chords intersect inside of a circle. The segments of one chord have lengths 3 and $x + 2$. The segments of the other chord have lengths x and $3x + 2$. What are the lengths of these chords?

a. 1 units
b. 2 units
c. 3 units
d. 6 units

23. Two chords intersect inside of a circle. The segments of one chord have the lengths 4 and $2x + 2$. The segments of the other chord have lengths x and $3x + 2$. What are the lengths of these chords?

a. 10 units
b. 2 units
c. 1 units
d. 3 units

24. Which of the following numbers has the greatest value?

a. 1.43785
b. 1.07548
c. 1.43592
d. 0.89409

25. The value of 6×12 is the same as:

a. $2 \times 4 \times 4 \times 2$
b. $7 \times 4 \times 3$
c. $6 \times 6 \times 3$
d. $3 \times 3 \times 4 \times 2$

26. This chart indicates how many sales of CDs, vinyl records, and MP3 downloads occurred over the last year. Approximately what percentage of the total sales was from CDs?

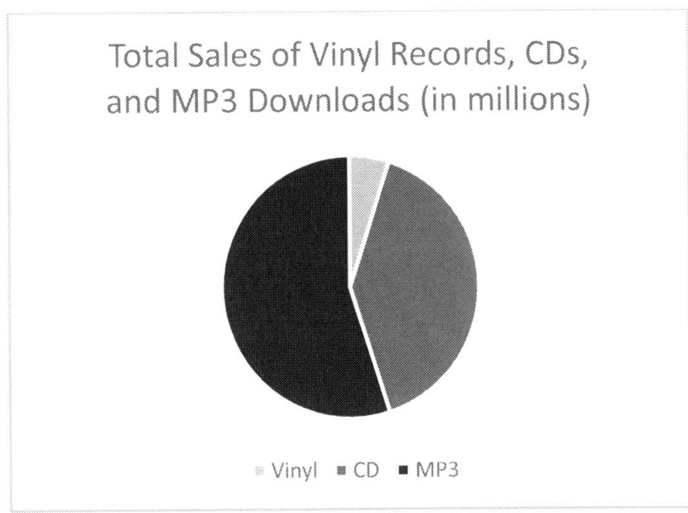

a. 55%
b. 25%
c. 40%
d. 5%

27. After a 20% sale discount, Frank purchased a new refrigerator for $850. How much did he save from the original price?
a. $170
b. $212.50
c. $105.75
d. $200

28. What is the value of b in this equation?

$$5b - 4 = 2b + 17$$

a. 13
b. 24
c. 7
d. 21

29. A school has 15 teachers and 20 teaching assistants. They have 200 students. What is the ratio of faculty to students?
a. $3:20$
b. $4:17$
c. $3:2$
d. $7:40$

30. Express the solution to the following problem in decimal form:

$$\frac{3}{5} \times \frac{7}{10} \div \frac{1}{2}$$

a. 0.042
b. 84%
c. 0.84
d. 0.42

31. Alan currently weighs 200 pounds, but he wants to lose weight to get down to 175 pounds. What is this difference in kilograms? (1 pound is approximately equal to 0.45 kilograms.)

a. 9 kg
b. 11.25 kg
c. 78.75 kg
d. 90 kg

32. Johnny earns $2,334.50 from his job each month. He pays $1,437 for monthly expenses. Johnny is planning a vacation in 3 months that he estimates will cost $1,750 total. How much will Johnny have left over from three months of saving once he pays for his vacation?

a. $948.50
b. $584.50
c. $852.50
d. $942.50

33. Solve the following:

$$4 \times 7 + (25 - 21)^2 \div 2$$

a. 512
b. 36
c. 60.5
d. 22

34. The total perimeter of a rectangle is 36 cm. If the length is 12 cm, what is the width?

a. 3 cm
b. 12 cm
c. 6 cm
d. 8 cm

35. Dwayne has received the following scores on his math tests: 78, 92, 83, 97. What score must Dwayne get on his next math test to have an overall average of 90?

a. 89
b. 98
c. 95
d. 100

36. What is the overall median of Dwayne's current scores: 78, 92, 83, 97?
   a. 19
   b. 85
   c. 83
   d. 87.5

37. In Jim's school, there are 3 girls for every 2 boys. There are 650 students in total. Using this information, how many students are girls?
   a. 260
   b. 130
   c. 65
   d. 390

38. Kimberley earns $10 an hour babysitting, and after 10 p.m., she earns $12 an hour, with the amount paid being rounded to the nearest hour accordingly. On her last job, she worked from 5:30 p.m. to 11 p.m. In total, how much did Kimberley earn on her last job?
   a. $45
   b. $57
   c. $62
   d. $42

39. What value of x would solve the following equation?

$$9x + x - 7 = 16 + 2x$$

   a. $x = -4$
   b. $x = 3$
   c. $x = \frac{9}{8}$
   d. $x = \frac{23}{8}$

40. Arrange the following numbers from least to greatest value:

$$0.85, \frac{4}{5}, \frac{2}{3}, \frac{91}{100}$$

   a. $0.85, \frac{4}{5}, \frac{2}{3}, \frac{91}{100}$
   b. $\frac{4}{5}, 0.85, \frac{91}{100}, \frac{2}{3}$
   c. $\frac{2}{3}, \frac{4}{5}, 0.85, \frac{91}{100}$
   d. $0.85, \frac{91}{100}, \frac{4}{5}, \frac{2}{3}$

41. Keith's bakery had 252 customers go through its doors last week. This week, that number increased to 378. Express this increase as a percentage.

   a. 26%
   b. 50%
   c. 35%
   d. 12%

42. If $4x - 3 = 5$, what is the value of x?

   a. 1
   b. 2
   c. 3
   d. 4

43. Simplify the following fraction:

$$\frac{\frac{5}{7}}{\frac{9}{11}}$$

   a. $\frac{55}{63}$
   b. $\frac{7}{1,000}$
   c. $\frac{13}{15}$
   d. $\frac{5}{11}$

44. The following graph compares the various test scores of the top three students in each of these teacher's classes. Based on the graph, which teacher's students had the lowest range of test scores?

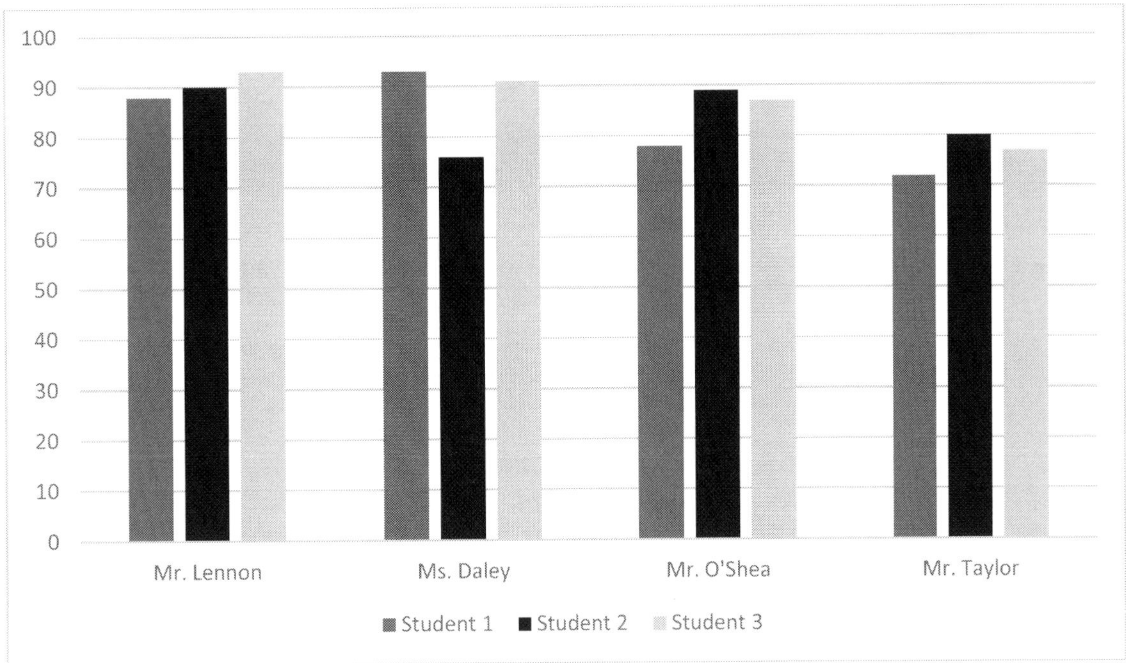

a. Mr. Lennon
b. Mr. O'Shea
c. Mr. Taylor
d. Ms. Daley

45. Bernard can make $80 per day. If he needs to make $300 and only works full days, how many days will this take?

a. 2
b. 3
c. 4
d. 5

46. Which measure for the center of a small sample set would be most affected by outliers?

a. Mean
b. Median
c. Mode
d. None of the above

Answer Explanations for Practice Test #2

1. A: Figure out which is largest by looking at the first non-zero digits. Choice *B*'s first non-zero digit is in the hundredths place. The other three all have non-zero digits in the tenths place, so it must be *A*, *C*, or *D*. Of these, *A* has the largest first non-zero digit.

2. C: $40N$ would be 4,000% of N. All of the other coefficients are equivalent to $\frac{40}{100}$ or 40%.

3. B: Instead of multiplying these out, the product can be estimated by using $18 \times 10 = 180$. The error here should be lower than 15, since it is rounded to the nearest integer, and the numbers add to something less than 30.

4. C: The average is calculated by adding all six numbers, then dividing by 6. The first five numbers have a sum of 25. If the total divided by 6 is equal to 6, then the total itself must be 36. The sixth number must be $36 - 25 = 11$.

5. B: $\frac{5}{2} \div \frac{1}{3} = \frac{5}{2} \times \frac{3}{1} = \frac{15}{2} = 7.5$.

6. A: The total fraction taken up by green and red shirts will be:

$$\frac{1}{3} + \frac{2}{5} = \frac{5}{15} + \frac{6}{15} = \frac{11}{15}$$

The remaining fraction is:

$$1 - \frac{11}{15} = \frac{15}{15} - \frac{11}{15} = \frac{4}{15}$$

7. C: If she has used $\frac{1}{3}$ of the paint, she has $\frac{2}{3}$ remaining. The mixed fraction can be converted because $2\frac{1}{2}$ gallons is the same as $\frac{5}{2}$ gallons. The calculation is:

$$\frac{2}{3} \times \frac{5}{2} = \frac{5}{3} = 1\frac{2}{3} \text{ gal}$$

8. A: If s is the size of the floor in square feet and r is the rate on Tuesday, then, based on the information given, $p = \frac{s}{4}$ and $r = \frac{s}{3}$. Solve the Monday rate for s, $s = 4p$, and then substitute that in the expression for Tuesday.

9. B: To be directly proportional means that $y = kx$. If x is changed from 5 to 20, the value of x is multiplied by 4. Applying the same rule to the y-value, also multiply the value of y by 4. Therefore:

$$y = 12$$

10. A: Each bag contributes $4x + 1$ treats. The total treats will be in the form $4nx + n$ where n is the total number of bags. The total is in the form $60x + 15$, from which it is known $n = 15$.

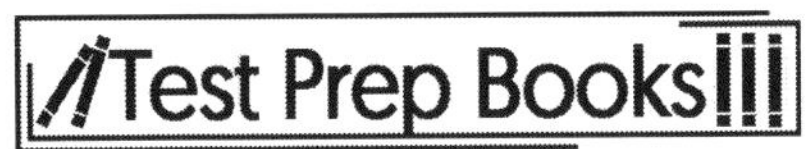

11. D: Denote the width as w and the length as l. Then, $l = 3w + 5$. The perimeter is $2w + 2l = 90$. Substituting the first expression for l into the second equation yields:

$$2(3w + 5) + 2w = 90$$

$$6w + 10 + 2w = 90$$

$$8w = 80$$

$$w = 10$$

Putting this into the first equation, it yields:

$$l = 3(10) + 5 = 35$$

12. B: If 60% of 50 workers are women, then there are 30 women working in the office. If half of them are wearing skirts, then that means 15 women wear skirts. Since none of the men wear skirts, this means there are 15 people wearing skirts.

13. B: The volume of a cube is the length of the side cubed, and 3 inches cubed is 27 in^3. Choice *A* is not the correct answer because that is 2×3 inches. Choice *C* is not the correct answer because that is 3×3 inches, and Choice *D* is not the correct answer because there was no operation performed.

14. D: The volume of a cube is the length of the side cubed, and 5 centimeters cubed is 125 cm^3. Choice *A* is not the correct answer because that is 2×5 centimeters. Choice *B* is not the correct answer because that is 3×5 centimeters. Choice *C* is not the correct answer because that is 5×10 centimeters.

15. B: Using the Pythagorean theorem, we can determine the length of the hypotenuse by plugging in the lengths of the sides $a^2 + b^2 = c^2$, or $3^2 + 4^2 = c^2$. We can then take the square root of 25 to get $c = 5$. Choice *A* is not the correct answer because that is $3 + 4$. Choice *C* is not the correct answer because that is stopping at $3^2 + 4^2$ is $9 + 16$, which is 25. Choice *D* is not the correct answer because that is 3×4.

16. A: We can use the Pythagorean theorem again to determine the length of the missing leg; i.e., $a^2 + b^2 = c^2$, or $a^2 + 8^2 = 10^2$. We then subtract $100 - 64 = a^2 = 36$ and then take the square root to give a length of 6. Choice *B* is not the correct answer because that is $10 + 8$. Choice *C* is not the correct answer because that is 8×10. Choice *D* is also not the correct answer because there is no reason to arrive at that number.

17. A: This answer is correct because $(2 + 2i)(2 - 2i)$, using the FOIL method and rules for imaginary numbers, is:

$$4 - 4i + 4i - 4i^2 = 8$$

Choice *B* is not the answer because there is no i in the final answer, since the i's cancel out in the FOIL. Choice *C*, 4, is not the final answer because we add $4 + 4$ in the end to equal 8. Choice *D*, $4i$, is not the final answer because there is neither a 4 nor an i in the final answer.

18. A: This answer is correct because $(3 + 3i)(3 - 3i)$, using the FOIL method and rules for imaginary numbers, is:

$$9 - 9i + 9i - 9i^2 = 18$$

Choice *B* is not the answer because there is no *i* in the final answer, since the *i*'s cancel out in the FOIL. Choice *C* is not the final answer because you have to add the two 9s together in the FOIL method. Choice *D* is not the final answer because there is neither a 9 nor an *i* in the final answer.

19. D: Multiply the top and the bottom by $(2 + 2i)$, the conjugate, to arrive at $\frac{8i}{8}$, which cancels to *i*. Choice *A* is not the answer because the 8's cancel out. Choice *B* is not the answer because the 8's cancel out. Choice *C* is not the answer because 2 is not left, but 8 is.

20. C: First, factor out the 3's: $\frac{1+i}{1-i}$. Then, multiply the top and bottom by its complex conjugate, $1 + i$:

$$\frac{(1 + i)(1 + i)}{(1 - i)(1 + i)} = \frac{1 + 2i + i^2}{1 - i^2}$$

Since i is the square root of -1, this goes to:

$$\frac{1 + 2i + (-1)}{1 - (-1)}$$

This equates to $\frac{2i}{2}$. Cancelling out the 2's leaves i. Choice *A* is not the correct answer because that only represents the denominator that is part of a fraction that needs to be simplified. Choice *B* is not the correct answer because that only represents the numerator that is part of a fraction that needs to be simplified. Choice *D* is not the final answer because it shows only part of the result from the FOIL method.

21. C: To find the angle in radians, multiply by pi and divide by 180. When you simplify $\frac{48° \times \pi}{180}$, you get $\frac{4\pi}{15}$. Choice *A* is not the correct answer because $\frac{\pi}{4}$ is 45°. Choice *B* is not the correct answer because $\frac{\pi}{15}$ is 12°. Choice *D* is not the correct answer because $\frac{3\pi}{4} = 135°$.

22. D: The method to equate the two chord lengths is $3 + x + 2 = x + 3x + 2$, add like terms, $5 + x = 4x + 2$, solve for x to yield $x = 1$, and substitute 1 back into the equation. Choice *A* is not the correct answer because 1 is the solution for x, not the length of the chord. Choice *B* is not the correct answer because 2 is one of the terms of the chord length when adding like terms. Choice *C* is not the correct answer because 3 is only the coefficient of one of the terms when solving.

23. A: The method is to equate the two chord lengths: $4 + 2x + 2 = x + 3x + 2$, add the like terms, $6 + 2x = 4x + 2$, solve for x to yield $x = 2$, and substitute 2 back into the equation. Choice *B* is not the correct answer because 2 is the solution for x, not the length of the chord. Choice *C* is not the correct answer because there is no 1 in the problem. Choice *D* is not the correct answer because 3 is only a coefficient in solving the equation.

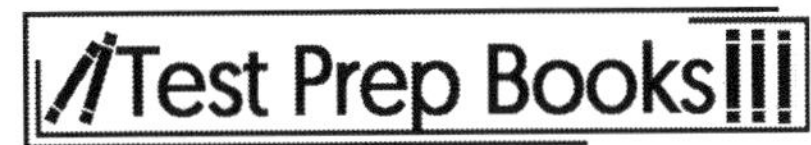

24. A: Compare each number after the decimal point to figure out which overall number is greatest. In Choices *A* (1.43785) and *C* (1.43592), both have the same tenths place (4) and hundredths place (3). However, the thousandths place is greater in Choice *A* (7), so *A* has the greatest value overall.

25. D: By grouping the four numbers in the answer into factors of the two numbers of the question (6 and 12), it can be determined that:

$$(3 \times 2) \times (4 \times 3) = 6 \times 12$$

Alternatively, you could find the prime factorization of each answer choices and compare it to the original value. The product of 6×12 is 72 and has a prime factorization of $2^3 \times 3^2$. The answer choices respectively have values of 64, 84, 108, and 72 and prime factorizations of 2^6, $2^2 \times 3 \times 7$, $2^2 \times 3^3$, and $2^3 \times 3^2$, so Choice *D* is the correct choice.

26. C: The sum total percentage of a pie chart must equal 100%. Since the CD sales take up less than half of the chart (50%) and more than a quarter (25%), it can be determined to be 40% overall. This can also be measured with a protractor. The angle of a circle is 360°. Since 25% of 360° would be 90° and 50% would be 180°, the angle percentage of CD sales falls in between; therefore, it would be Choice *C*.

27. B: Since $850 is the price *after* a 20% discount, $850 represents 80% of the original price. To determine the original price, set up a proportion with the ratio of the sale price (850) to the original price (unknown) equal to the ratio of the sale percentage (where x represents the unknown original price):

$$\frac{850}{x} = \frac{80}{100}$$

To solve a proportion, cross multiply and set the products equal to each other:

$$(850)(100) = (80)(x)$$

Multiplying each side results in the equation:

$$85{,}000 = 80x$$

To solve for x, divide both sides by 80:

$$\frac{85{,}000}{80} = \frac{80x}{80}$$

$$x = 1{,}062.5$$

Remember that x represents the original price. Subtracting the sale price from the original price ($1,062.50 – $850) indicates that Frank saved $212.50.

28. C: To solve for the value of b, isolate the variable b on one side of the equation.

Start by moving the lower value of -4 to the other side by adding 4 to both sides:

$$5b - 4 = 2b + 17$$

$$5b - 4 + 4 = 2b + 17 + 4$$

$$5b = 2b + 21$$

Then subtract $2b$ from both sides:

$$5b - 2b = 2b + 21 - 2b$$

$$3b = 21$$

Then divide both sides by 3 to get the value of b:

$$\frac{3b}{3} = \frac{21}{3}$$

$$b = 7$$

29. D: The total faculty is:

$$15 + 20 = 35$$

So, the ratio is $35 : 200$. Then, divide both of these numbers by 5, since 5 is a common factor to both, with a result of $7 : 40$.

30. C: The first step in solving this problem is expressing the result in fraction form. Multiplication and division are typically performed in order from left to right but they can be performed in any order. For this problem, let's start by solving the division operation of the last two fractions. When dividing one fraction by another, invert or flip the second fraction and then multiply the numerator and denominator.

$$\frac{7}{10} \times \frac{2}{1} = \frac{14}{10}$$

Next, multiply the first fraction with this value:

$$\frac{3}{5} \times \frac{14}{10} = \frac{42}{50}$$

In this instance, you can find the decimal form by converting the fraction into $\frac{x}{100}$, where x is the number from which the final decimal is found. Multiply both the numerator and denominator by 2 to get the fraction as an expression of $\frac{x}{100}$.

$$\frac{42}{50} \times \frac{2}{2} = \frac{84}{100}$$

In decimal form, this would be expressed as 0.84.

31. B: Using the conversion rate, multiply the projected weight loss of 25 lb by 0.45 kg/lb to get the amount in kilograms (11.25 kg).

32. D: First, subtract $1,437 from $2,334.50 to find Johnny's monthly savings; this equals $897.50. Then, multiply this amount by 3 to find out how much he will have (in three months) before he pays for his vacation: this equals $2,692.50. Finally, subtract the cost of the vacation ($1,750) from this amount to find how much Johnny will have left: $942.50.

33. B: To solve this correctly, keep in mind the order of operations with the mnemonic PEMDAS (Please Excuse My Dear Aunt Sally). This stands for Parentheses, Exponents, Multiplication, Division, Addition, Subtraction. Taking it step by step, solve inside the parentheses first:

$$4 \times 7 + (4)^2 \div 2$$

Then, apply the exponent:

$$4 \times 7 + 16 \div 2$$

Multiplication and division are both performed next:

$$28 + 8$$

And then, addition:

$$28 + 8 = 36$$

The solution is 36.

34. C: The formula for the perimeter of a rectangle is $P = 2L + 2W$, where P is the perimeter, L is the length, and W is the width. The first step is to substitute all of the data into the formula:

$$36 = 2(12) + 2W$$

Simplify by multiplying 2×12:

$$36 = 24 + 2W$$

Simplifying this further by subtracting 24 on each side gives:

$$36 - 24 = 24 - 24 + 2W$$

$$12 = 2W$$

Divide by 2:

$$6 = W$$

The width is 6 cm. Remember to test this answer by substituting this value into the original formula:

$$36 = 2(12) + 2(6)$$

35. D: To find the average of a set of values, add the values together and then divide by the total number of values. In this case, include the unknown value of what Dwayne needs to score on his next test, in order to solve it.

$$\frac{78 + 92 + 83 + 97 + x}{5} = 90$$

Add the unknown value to the new average total, which is 5. Then multiply each side by 5 to simplify the equation, resulting in:

$$78 + 92 + 83 + 97 + x = 450$$

$$350 + x = 450$$

$$x = 100$$

Dwayne would need to get a perfect score of 100 in order to get an average of at least 90.

Test this answer by substituting back into the original formula.

$$\frac{78 + 92 + 83 + 97 + 100}{5} = 90$$

36. D: For an even number of total values, the *median* is calculated by finding the *mean,* or average, of the two middle values once all values have been arranged in ascending order from least to greatest. In this case, $(92 + 83) \div 2$ would equal the median 87.5, Choice *D*.

37. D: Three girls for every two boys can be expressed as a ratio: 3: 2. This can be visualized as splitting the school into 5 groups: 3 girl groups and 2 boy groups. The number of students that are in each group can be found by dividing the total number of students by 5:

$$\frac{650 \text{ students}}{5 \text{ groups}} = \frac{130 \text{ students}}{\text{group}}$$

To find the total number of girls, multiply the number of students per group (130) by the number of girl groups in the school (3). This equals 390, Choice *D*.

38. C: Kimberley worked 4.5 hours at the rate of \$10/h and 1 hour at the rate of \$12/h. The problem states that her pay is rounded to the nearest hour, so the 4.5 hours would round up to 5 hours at the rate of \$10/h.

$$(5\text{h}) \times \left(\frac{\$10}{\text{h}}\right) + (1 \text{ h}) \times \left(\frac{\$12}{\text{h}}\right) = \$50 + \$12 = \$62$$

39. D:

$9x + x - 7 = 16 + 2x$	Combine $9x$ and x.
$10x - 7 = 16 + 2x$	
$10x - 7 + 7 = 16 + 2x + 7$	Add 7 to both sides to remove (-7).
$10x = 23 + 2x$	
$10x - 2x = 23 + 2x - 2x$	Subtract $2x$ from both sides to move it to the other side of the equation.
$8x = 23$	
$\frac{8x}{8} = \frac{23}{8}$	Divide by 8 to get x by itself.
$x = \frac{23}{8}$	

40. C: The first step is to depict each number using decimals:

$$\frac{91}{100} = 0.91$$

Dividing the numerator by the denominator of $\frac{4}{5}$ to convert it to a decimal yields 0.80, while $\frac{2}{3}$ becomes 0.66 recurring. Rearrange each expression in ascending order, as found in Choice *C*.

41. B: First, calculate the difference between the larger value and the smaller value:

$$378 - 252 = 126$$

To calculate this difference as a percentage of the original value, and thus calculate the percentage *increase*, divide 126 by 252, then multiply by 100 to reach the percentage 50%, Choice *B*.

42. B: Add 3 to both sides to get $4x = 8$. Then divide both sides by 4 to get $x = 2$.

43. A: First, simplify the larger fraction by separating it into two. When dividing one fraction by another, remember to *invert* the second fraction and multiply the two as follows:

$$\frac{5}{7} \times \frac{11}{9}$$

The resulting fraction $\frac{55}{63}$ cannot be simplified further, so this is the answer to the problem.

44. A: To calculate the range in a set of data, subtract the lowest value from the highest value. In this graph, the range of Mr. Lennon's students is 4, which can be seen physically in the graph as having the smallest difference compared with the other teachers between the highest value and the lowest value.

45. C: The number of days can be found by taking the total amount Bernard needs to make and dividing it by the amount he earns per day:

$$\frac{300}{80} = \frac{30}{8} = \frac{15}{4} = 3.75$$

But Bernard is only working full days, so he will need to work 4 days since 3 days is not a sufficient amount of time.

46. A: Mean. An outlier is a data value that's either far above or below the majority of values in a sample set. The mean is the average of all values in the set. In a small sample, a very high or low number could greatly change the average. The median is the middle value when arranged from lowest to highest. Outliers would have no more of an effect on the median than any other value. Mode is the value that repeats most often in a set. Assuming that the same outlier doesn't repeat, outliers would have no effect on the mode of a sample set.

GED Mathematical Reasoning Practice Test #3

1. $5.88 \times 3.2 =$
   a. 18.816
   b. 16.44
   c. 20.352
   d. 17

2. How will the following number be written in standard form:

$$(1 \times 10^4) + (3 \times 10^3) + (7 \times 10^1) + (8 \times 10^0)$$

   a. 137
   b. 13,078
   c. 1,378
   d. 8,731

3. What is the value of the expression: $7^2 - 3 \times (4 + 2) + 15 \div 5$?
   a. 12.2
   b. 40.2
   c. 34
   d. 58.2

4. Four people split a bill. The first person pays for $\frac{1}{5}$, the second person pays for $\frac{1}{4}$, and the third person pays for $\frac{1}{3}$. What fraction of the bill does the fourth person pay?
   a. $\frac{13}{60}$
   b. $\frac{47}{60}$
   c. $\frac{1}{4}$
   d. $\frac{4}{15}$

5. A student gets an 85% on a test with 20 questions. How many answers did the student solve correctly?
   a. 15
   b. 16
   c. 17
   d. 18

6. What is $\frac{420}{98}$ rounded to the nearest integer?
   a. 4
   b. 3
   c. 5
   d. 6

7. If Danny takes 48 minutes to walk 3 miles, how long should it take him to walk 5 miles maintaining the same speed?
   a. 32 min
   b. 64 min
   c. 80 min
   d. 96 min

8. If $\sqrt{1+x} = 4$, what is x?
   a. 10
   b. 15
   c. 20
   d. 25

9. $52.3 \times 10^{-3} =$
   a. 0.00523
   b. 0.0523
   c. 0.523
   d. 523

10. Which of the following is a factor of both $x^2 + 4x + 4$ and $x^2 - x - 6$?
   a. $x - 3$
   b. $x + 2$
   c. $x - 2$
   d. $x + 3$

11. What is the simplified form of the expression $(7n + 3n^3 + 3) + (8n + 5n^3 + 2n^4)$?
   a. $9n^4 + 15n - 2$
   b. $2n^4 + 5n^3 + 15n - 2$
   c. $9n^4 + 8n^3 + 15n$
   d. $2n^4 + 8n^3 + 15n + 3$

12. Which of the following inequalities is equivalent to $3 - \frac{1}{2}x \geq 2$?
   a. $x \geq 2$
   b. $x \leq 2$
   c. $x \geq 1$
   d. $x \leq 1$

13. For which of the following are $x = 4$ and $x = -4$ solutions?
   a. $x^2 + 16 = 0$
   b. $x^2 + 4x - 4 = 0$
   c. $x^2 - 2x - 2 = 0$
   d. $x^2 - 16 = 0$

14. What is the solution to the following system of equations?

$$x^2 - 2x + y = 8$$
$$x - y = -2$$

a. $(-2, 3)$
b. There is no solution.
c. $(-2, 0)$ and $(1, 3)$
d. $(-2, 0)$ and $(3, 5)$

15. A line passes through the point $(1, 2)$ and crosses the y-axis at $y = 1$. Which of the following is an equation for this line?

a. $y = 2x$
b. $y = x + 1$
c. $x + y = 1$
d. $y = \frac{x}{2} - 2$

16. A company invests $50,000 in a building where they can produce saws. If the cost of producing one saw is $40, then which function expresses the amount of money the company pays? The variable y is the money paid and x is the number of saws produced.

a. $y = 50{,}000x + 40$
b. $y + 40 = x - 50{,}000$
c. $y = 40x - 50{,}000$
d. $y = 40x + 50{,}000$

17. If $x > 3$, then $\frac{x^2-6x+9}{x^2-x-6} =$

a. $\frac{x+2}{x-3}$
b. $\frac{x-2}{x-3}$
c. $\frac{x-3}{x+3}$
d. $\frac{x-3}{x+2}$

18. Is the following function even, odd, neither, or both?

$$y = \frac{1}{2}x^4 + 2x^2 - 6$$

a. Even
b. Odd
c. Neither
d. Both

19. Christie is building a shed with a base of 7 feet by 3.5 feet. If she plans to make the walls seven feet tall, how many 7-foot-long, 3.5-inch-wide boards will it take to completely surround the base, if there is no overlap?

a. 36
b. 48
c. 72
d. 108

20. What is the 42nd item in the pattern: ▲oo□▲oo□▲...?
    a. o
    b. ▲
    c. □
    d. None of the above

21. For a group of 20 men, the median weight is 180 pounds and the range is 30 pounds. If each man gains 10 pounds, which of the following would be true?
    a. The median weight will increase, and the range will remain the same.
    b. The median weight and range will both remain the same.
    c. The median weight will stay the same, and the range will increase.
    d. The median weight and range will both increase.

22. Five students take a test. The scores of the first four students are 80, 85, 75, and 60. If the median score is 80, which of the following could NOT be the score of the fifth student?
    a. 60
    b. 80
    c. 85
    d. 100

23. Ten students take a test. Five students get a 50. Four students get a 70. If the average score is 55, what was the last student's score?
    a. 20
    b. 40
    c. 50
    d. 60

24. Given the value of a given stock at monthly intervals, which graph should be used to best represent the trend of the stock?
    a. Box plot
    b. Line plot
    c. Line graph
    d. Circle graph

25. A six-sided die is rolled. What is the probability that the roll is 1 or 2?
    a. $\frac{1}{6}$
    b. $\frac{1}{4}$
    c. $\frac{1}{3}$
    d. $\frac{1}{2}$

26. What is the probability of randomly picking the winner and runner-up from a race of four horses and distinguishing which is the winner?
    a. $\frac{1}{4}$
    b. $\frac{1}{2}$
    c. $\frac{1}{16}$
    d. $\frac{1}{12}$

27. A grocery store is selling individual bottles of water, and each bottle contains 750 milliliters of water. If 12 bottles are purchased, what conversion will correctly determine how many liters that customer will take home?
    a. 100 milliliters equals 1 liter
    b. 1,000 milliliters equals 1 liter
    c. 1,000 liters equals 1 milliliter
    d. 10 liters equals 1 milliliter

28. Which of the following statements is true about the two lines below?

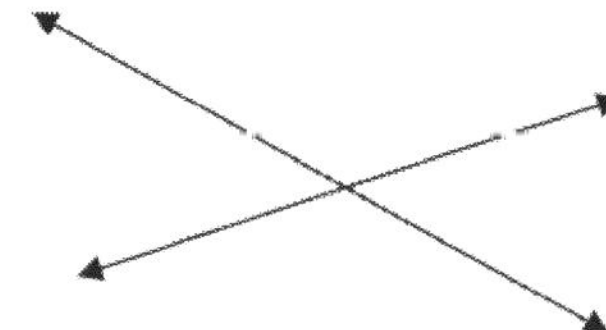

    a. The two lines are parallel but not perpendicular.
    b. The two lines are perpendicular but not parallel.
    c. The two lines are both parallel and perpendicular.
    d. The two lines are neither parallel nor perpendicular.

29. The perimeter of a 6-sided polygon is 56 cm. The lengths of three sides are 9 cm each. The lengths of two other sides are 8 cm each. What is the length of the missing side?
    a. 11 cm
    b. 12 cm
    c. 13 cm
    d. 10 cm

30. An equilateral triangle has a perimeter of 18 feet. If a square whose sides have the same length as one side of the triangle is built, what will be the area of the square?
    a. 6 square feet
    b. 36 square feet
    c. 256 square feet
    d. 1,000 square feet

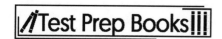

31. The area of a given rectangle is 24 square centimeters. If the measure of each side is multiplied by 3, what is the area of the new figure?

a. $48\ cm^2$
b. $72\ cm^2$
c. $216\ cm^2$
d. $13{,}824\ cm^2$

32. Apples cost $2 each, while bananas cost $3 each. Maria purchased 10 fruits in total and spent $22. How many apples did she buy?

a. 5
b. 6
c. 7
d. 8

33. $(4x^2y^4)^{\frac{3}{2}}$ can be simplified to which of the following?

a. $8x^3y^6$
b. $4x^{\frac{5}{2}}y$
c. $4xy$
d. $32x^{\frac{7}{2}}y^{\frac{11}{2}}$

34. A line passes through the origin and through the point $(-3, 4)$. What is the slope of the line?

a. $-\frac{4}{3}$
b. $-\frac{3}{4}$
c. $\frac{4}{3}$
d. $\frac{3}{4}$

35. $3\frac{2}{3} - 1\frac{4}{5} =$

a. $1\frac{13}{15}$
b. $\frac{14}{15}$
c. $2\frac{2}{3}$
d. $\frac{4}{5}$

36. What is the value of $x^2 - 2xy + 2y^2$ when $x = 2, y = 3$?

a. 8
b. 10
c. 12
d. 14

37. The square and circle have the same center. The circle has a radius of r. What is the area of the shaded region?

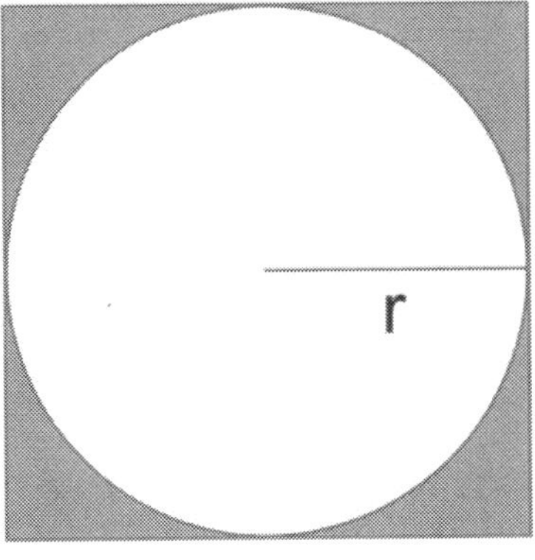

a. $r^2 - \pi r^2$
b. $4r^2 - 2\pi r$
c. $(4 - \pi)r^2$
d. $(\pi - 1)r^2$

38. What is the equation for the line passing through the origin and the point $(2,1)$?

a. $y = 2x$
b. $y = \frac{1}{2}x$
c. $y = x - 2$
d. $2y = x + 1$

39. What are the coordinates of the two points marked with dots on this coordinate plane?

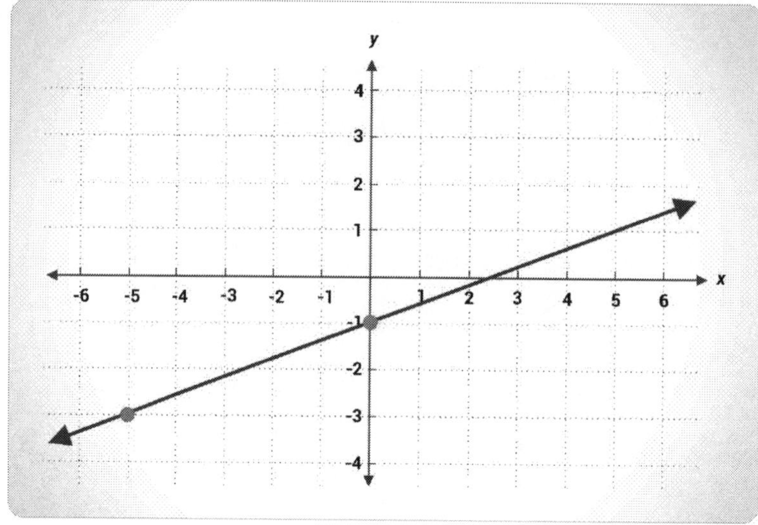

a. $(-3, -5)$ and $(-1, 0)$
b. $(5, 3)$ and $(0, 1)$
c. $(-5, -3)$ and $(0, -1)$
d. $(-3, -5)$ and $(0, -1)$

40. What is the value of x for the right triangle shown below?

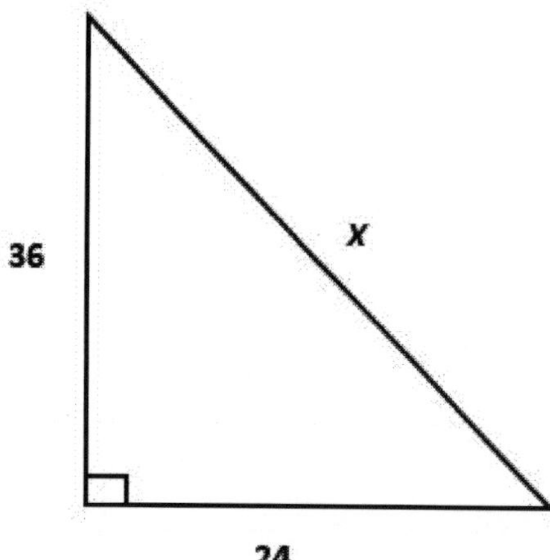

a. 43.3
b. 26.8
c. 42.7
d. 44.1

41. A cube has sides that are 7 inches long. What is the cube's volume?
   a. $49in^3$
   b. $343in^3$
   c. $294in^3$
   d. $28in^3$

42. $3.4 + 2.35 + 4 =$
   a. 5.35
   b. 9.2
   c. 9.75
   d. 10.25

43. $\frac{3}{25} =$
   a. 0.15
   b. 0.1
   c. 0.9
   d. 0.12

44. 6 is 30% of what number?
   a. 18
   b. 20
   c. 24
   d. 26

45. What is the value of the following expression?

$$\sqrt{8^2 + 6^2}$$

   a. 14
   b. 10
   c. 9
   d. 100

46. $864 \div 36 =$
   a. 24
   b. 25
   c. 34
   d. 18

Answer Explanations for Practice Test #3

1. A: This problem can be multiplied as 588×32, except at the end, the decimal point needs to be moved three places to the left. Performing the multiplication will give 18,816, and moving the decimal place over three places results in 18.816.

2. B: 13,078. The power of 10 by which a digit is multiplied corresponds with the number of zeros following the digit when expressing its value in standard form. Therefore,

$$(1 \times 10^4) + (3 \times 10^3) + (7 \times 10^1) + (8 \times 10^0)$$

$$10{,}000 + 3{,}000 + 70 + 8 = 13{,}078$$

3. C: When performing calculations consisting of more than one operation, the order of operations should be followed: Parentheses, Exponents, Multiplication/Division, Addition/Subtraction.

Parentheses:

$$7^2 - 3 \times (4 + 2) + 15 \div 5$$

$$7^2 - 3 \times (6) + 15 \div 5$$

Exponents:

$$49 - 3 \times 6 + 15 \div 5$$

Multiplication/Division (from left to right):

$$49 - 18 + 3$$

Addition/Subtraction (from left to right):

$$49 - 18 + 3 = 34$$

4. A: To find the fraction of the bill that the first three people pay, the fractions need to be added, which means finding the common denominator. The common denominator will be 60.

$$\frac{1}{5} + \frac{1}{4} + \frac{1}{3} = \frac{12}{60} + \frac{15}{60} + \frac{20}{60} = \frac{47}{60}$$

The remainder of the bill is:

$$1 - \frac{47}{60} = \frac{60}{60} - \frac{47}{60} = \frac{13}{60}$$

5. C: 85% of a number means multiplying that number by 0.85. So, $0.85 \times 20 = \frac{85}{100} \times \frac{20}{1}$, which can be simplified to:

$$\frac{17}{20} \times \frac{20}{1} = 17$$

Therefore, the student got 17 questions correct.

6. A: Dividing by 98 can be approximated by dividing by 100, which would mean shifting the decimal point of the numerator to the left by 2. The result is 4.2 and rounds to 4.

7. C: To solve the problem, a proportion is written consisting of ratios comparing distance and time. One way to set up the proportion is: $\frac{3}{48} = \frac{5}{x} \left(\frac{distance}{time} = \frac{distance}{time}\right)$ where *x* represents the unknown value of time. To solve a proportion, the ratios are cross-multiplied:

$$(3)(x) = (5)(48) \rightarrow 3x = 240$$

The equation is solved by isolating the variable, or dividing by 3 on both sides, to produce $x = 80$.

8. B: Start by squaring both sides to get $1 + x = 16$. Then subtract 1 from both sides to get $x = 15$.

9. B: Multiplying by 10^{-3} means moving the decimal point three places to the left, putting in zeros as necessary.

10. B: To factor $x^2 + 4x + 4$, the numbers needed are those that add to 4 and multiply to 4. Therefore, both numbers must be 2, and the expression factors to:

$$x^2 + 4x + 4 = (x + 2)^2$$

Similarly, the second expression factors to $x^2 - x - 6 = (x - 3)(x + 2)$, so that they have $x + 2$ in common.

11. D: The expression is simplified by collecting like terms. Terms with the same variable and exponent are like terms, and their coefficients can be added.

12. B: To simplify this inequality, subtract 3 from both sides to get $-\frac{1}{2}x \geq -1$. Then, multiply both sides by -2 (remembering this flips the direction of the inequality) to get $x \leq 2$.

13. D: There are two ways to approach this problem. Each value can be substituted into each equation. A can be eliminated, since:

$$4^2 + 16 = 32$$

Choice *B* can be eliminated, since:

$$4^2 + 4 \cdot 4 - 4 = 28$$

C can be eliminated, since:

$$4^2 - 2 \cdot 4 - 2 = 6$$

But, plugging in either value into $x^2 - 16$, which gives:

$$(\pm 4)^2 - 16 = 16 - 16 = 0$$

14. D: This system of equations involves one quadratic function and one linear function, as seen from the degree of each equation. One way to solve this is through substitution.

Solving for y in the second equation yields:

$$y = x + 2$$

Plugging this equation in for the y of the quadratic equation yields:

$$x^2 - 2x + x + 2 = 8$$

Simplifying the equation, it becomes:

$$x^2 - x + 2 = 8$$

Setting this equal to zero and factoring, it becomes:

$$x^2 - x - 6 = 0 = (x - 3)(x + 2)$$

Solving these two factors for x gives the zeros:

$$x = 3, -2$$

To find the y-value for the point, each number can be plugged in to either original equation. Solving each one for y yields the points $(3, 5)$ and $(-2, 0)$.

15. B: From the slope-intercept form, $y = mx + b$, it is known that b is the y-intercept, which is 1. Compute the slope as $\frac{2-1}{1-0} = 1$, so the equation is $y = x + 1$.

16. D: For manufacturing costs, there is a linear relationship between the cost to the company and the number produced, with a y-intercept given by the base cost of acquiring the means of production and a slope given by the cost to produce one unit. In this case, that base cost is $50,000, while the cost per unit is $40. So:

$$y = 40x + 50{,}000$$

17. D: Factor the numerator into $x^2 - 6x + 9 = (x - 3)^2$, since $-3 - 3 = -6, (-3)(-3) = 9$. Factor the denominator into $x^2 - x - 6 = (x - 3)(x + 2)$, since $-3 + 2 = -1, (-3)(2) = -6$. This means the rational function can be rewritten as:

$$\frac{x^2 - 6x + 9}{x^2 - x - 6} = \frac{(x - 3)^2}{(x - 3)(x + 2)}$$

Using the restriction of $x > 3$, do not worry about any of these terms being 0, and cancel an $x - 3$ from the numerator and the denominator, leaving $\frac{x-3}{x+2}$.

18. A: The equation is *even* because $f(-x) = f(x)$. Plugging in a negative value will result in the same answer as when plugging in the positive of that same value. The function:

$$f(-2) = \frac{1}{2}(-2)^4 + 2(-2)^2 - 6 = 8 + 8 - 6 = 10$$

This yields the same value as:

$$f(2) = \frac{1}{2}(2)^4 + 2(2)^2 - 6 = 8 + 8 - 6 = 10$$

19. C: Christie would need 72 boards. The total can be found by converting the length and width of the base into inches and finding the number of boards for each side. Start by finding the dimensions in inches:

$$7 \text{ ft} \times \frac{12 \text{ in}}{1 \text{ ft}} = 84 \text{ in}$$

$$3.5 \text{ ft} \times \frac{12 \text{ in}}{1 \text{ ft}} = 42 \text{ in}$$

Next, divide the length of each side by 3.5 inches per board to find the number of boards needed to cover that side, and multiply by 2 to account for the opposite walls:

$$84 \text{ in} \times \frac{1 \text{ board}}{3.5 \text{ in}} \times 2 = 48 \text{ boards}$$

$$42 \text{ in} \times \frac{1 \text{ board}}{3.5 \text{ in}} \times 2 = 24 \text{ boards}$$

Finally, sum the numbers for both lengths and widths to arrive at the total number of boards needed:

$$48 \text{ boards} + 24 \text{ boards} = 72 \text{ boards}$$

20. A: ○. The core of the pattern consists of 4 items: ▲○○□. Therefore, the core repeats in multiples of 4, with the pattern starting over on the next step. The closest multiple of 4 to 42 is 40. Step 40 is the end of the core (□), so step 41 will start the core over (▲) and step 42 is ○.

21. A: If each man gains 10 pounds, every original data point will increase by 10 pounds. Therefore, the man with the original median will still have the median value, but that value will increase by 10. The smallest value and largest value will also increase by 10 and, therefore, the difference between the two won't change. The range does not change in value and, thus, remains the same.

22. A: Lining up the given scores provides the following list: 60, 75, 80, 85, and one unknown. Because the median needs to be 80, it means 80 must be the middle data point out of these five. Therefore, the unknown data point must be the fourth or fifth data point, meaning it must be greater than or equal to 80. The only answer that fails to meet this condition is 60.

23. A: Let the unknown score be x. The average will be:

$$\frac{5 \times 50 + 4 \times 70 + x}{10} = \frac{530 + x}{10} = 55$$

Multiply both sides by 10 to get $530 + x = 550$, or $x = 20$.

24. C: Line graph. The scenario involves data consisting of two variables, month and stock value. Box plots display data consisting of values for one variable. Therefore, a box plot is not an appropriate choice. Both line plots and circle graphs are used to display frequencies within categorical data. Neither can be used for the given scenario. Line graphs display two numerical variables on a coordinate grid and show trends among the variables.

25. C: A die has an equal chance for each outcome. Since it has six sides, each outcome has a probability of $\frac{1}{6}$. The chance of a 1 or a 2 is therefore $\frac{1}{6}+\frac{1}{6}=\frac{1}{3}$.

26. D: $\frac{1}{12}$. The probability of picking the winner of the race is $\frac{1}{4}$, or $\left(\frac{number\ of\ favorable\ outcomes}{number\ of\ total\ outcomes}\right)$. Assuming the winner was picked on the first selection, three horses remain from which to choose the runner-up (these are dependent events). Therefore, the probability of picking the runner-up is $\frac{1}{3}$. To determine the probability of multiple events, the probability of each event is multiplied:

$$\frac{1}{4}\times\frac{1}{3}=\frac{1}{12}$$

27. B: $12 \times 750 = 9{,}000$. Therefore, there are 9,000 milliliters of water, which must be converted to liters. 1,000 milliliters equals 1 liter; therefore, 9 liters of water are purchased.

28. D: The two lines are neither parallel nor perpendicular. Parallel lines will never intersect or meet. Therefore, the lines are not parallel. Perpendicular lines intersect to form a right angle (90°). Although the lines intersect, they do not form a right angle, which is usually indicated with a box at the intersection point. Therefore, the lines are not perpendicular.

29. C: Perimeter is found by calculating the sum of all sides of the polygon. $9+9+9+8+8+s=56$, where s is the missing side length. Therefore, $43+s=56$. The missing side length is 13 cm.

30. B: An equilateral triangle has three sides of equal length, so if the total perimeter is 18 feet, each side must be 6 feet long. A square with sides of 6 feet will have an area of $6^2 = 36$ square feet.

31. C: Because area is a two-dimensional measurement, the dimensions are multiplied by a scale factor that is squared to determine the scale factor of the corresponding areas. The dimensions of the rectangle are multiplied by a scale factor of 3. Therefore, the area is multiplied by a scale factor of 3^2 (which is equal to 9):

$$24\text{ cm}^2 \times 9 = 216\text{ cm}^2$$

32. D: Let a be the number of apples and b be the number of bananas. Then, the total cost is:

$$2a+3b=22$$

It also known that:

$$a+b=10$$

Using the knowledge of systems of equations, cancel the b-variables by multiplying the second equation by -3. This makes the equation:

$$-3a-3b=-30$$

Adding this to the first equation, the b-values cancel to get $-a=-8$, which simplifies to $a=8$.

33. A: Simplify this to:

$$(4x^2y^4)^{\frac{3}{2}} = 4^{\frac{3}{2}}(x^2)^{\frac{3}{2}}(y^4)^{\frac{3}{2}}$$

Now,

$$4^{\frac{3}{2}} = (\sqrt{4})^3 = 2^3 = 8$$

For the other, recall that the exponents must be multiplied, so this yields:

$$8x^{2\cdot\frac{3}{2}}y^{4\cdot\frac{3}{2}} = 8x^3y^6$$

34. A: The slope is given by:

$$m = \frac{y_2 - y_1}{x_2 - x_1} = \frac{0 - 4}{0 - (-3)} = -\frac{4}{3}$$

35. A: First, convert the mixed numbers to improper fractions: $\frac{11}{3} - \frac{9}{5}$. Then, convert the fractions to have a common denominator of 15, subtract, and convert the answer back to a mixed number.

$$\frac{11}{3} - \frac{9}{5} = \frac{55}{15} - \frac{27}{15} = \frac{28}{15} = 1\frac{13}{15}$$

36. B: Start with the original equation: $x^2 - 2xy + 2y$, then replace each instance of x with a 2, and each instance of y with a 3 to get:

$$2^2 - 2 \times 2 \times 3 + 2 \times 3^2 = 4 - 12 + 18 = 10$$

37. C: The area of the shaded region is the area of the square minus the area of the circle. The area of the circle is πr^2. The side of the square will be $2r$, so the area of the square will be $4r^2$. Therefore, the difference is:

$$4r^2 - \pi r^2 = (4 - \pi)r^2$$

38. B: The slope will be given by:

$$m = \frac{y_2 - y_1}{x_2 - x_1}$$

$$m = \frac{1 - 0}{2 - 0} = \frac{1}{2}$$

The y-intercept will be 0 since it passes through the origin. Using slope-intercept form, the equation for this line is:

$$y = \frac{1}{2}x$$

39: C. The two points are at -5 and 0 for the x-axis and at -3 and at -1 for y-axis respectively. Therefore, the two points have the coordinates of $(-5, -3)$ and $(0, -1)$.

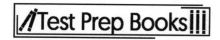

40. A: The Pythagorean theorem states that for right triangles $c^2 = a^2 + b^2$, with c being the side opposite the 90° angle. Substituting 24 as a and 36 as b, the equation becomes:

$$c^2 = 24^2 + 36^2 = 576 + 1296 = 1872$$

The last step is to square both sides to remove the exponent:

$$c = \sqrt{1872} = 43.3$$

41. B: The formula for the volume of a cube is $V = s^3$. Substitute the side length of 7 in to get:

$$V = 7^3 = 343in^3$$

42. C: The decimal points are lined up, with zeroes put in as needed. Then, the numbers are added just like integers:

$$\begin{array}{r} 3.40 \\ 2.35 \\ +\underline{4.00} \\ 9.75 \end{array}$$

43. D: The fraction is converted so that the denominator is 100 by multiplying the numerator and denominator by 4, to get $\frac{3}{25} = \frac{12}{100}$. Dividing a number by 100 just moves the decimal point two places to the left, with a result of 0.12.

44. B: 30% is $\frac{3}{10}$. The equation to represent this question is $6 = \frac{3}{10}x$. To find the value of x, multiply both sides by $\frac{3}{10}$. This results in $20 = x$, so 20 is 30% of 6.

$$\frac{10}{3} \times 6 = 10 \times 2 = 20$$

45. B: 8 squared is 64, and 6 squared is 36. These should be added together to get $64 + 36 = 100$. Then, the last step is to find the square root of 100 which is 10.

46. A: The long division would be completed as follows:

$$\begin{array}{r} 24 \\ 36\overline{)864} \\ -\underline{72\downarrow} \\ 144 \end{array}$$

Index

Dear GED Math Test Taker,

We would like to start by thanking you for purchasing this study guide for the math section of your GED exam. We hope that we exceeded your expectations.

Our goal in creating this study guide was to cover all of the topics that you will see on the math section of the test. We also strove to make our practice questions as similar as possible to what you will encounter on test day. With that being said, if you found something that you feel was not up to your standards, please send us an email and let us know.

We would also like to let you know about other books in our catalog that may interest you.

HiSET

amazon.com/dp/1637753357

SAT

amazon.com/dp/1637759878

ACT

amazon.com/dp/163775583X

ACCUPLACER

amazon.com/dp/1637750250

We have study guides in a wide variety of fields. If the one you are looking for isn't listed above, then try searching for it on Amazon or send us an email.

Thanks Again and Happy Testing!
Product Development Team
info@studyguideteam.com

FREE Test Taking Tips Video/DVD Offer

To better serve you, we created videos covering test taking tips that we want to give you for FREE. **These videos cover world-class tips that will help you succeed on your test.**

We just ask that you send us feedback about this product. Please let us know what you thought about it—whether good, bad, or indifferent.

To get your **FREE videos**, you can use the QR code below or email freevideos@studyguideteam.com with "Free Videos" in the subject line and the following information in the body of the email:

a. The title of your product

b. Your product rating on a scale of 1-5, with 5 being the highest

c. Your feedback about the product

If you have any questions or concerns, please don't hesitate to contact us at info@studyguideteam.com.

Thank you!

Made in the USA
Monee, IL
21 January 2023